Streaming Video Collection Development and Management

Streaming Video Collection Development and Management

Michael Fernandez and Amauri Serrano

BLOOMSBURY LIBRARIES UNLIMITED
NEW YORK • LONDON • OXFORD • NEW DELHI • SYDNEY

BLOOMSBURY LIBRARIES UNLIMITED
Bloomsbury Publishing Inc
1385 Broadway, New York, NY 10018, USA
50 Bedford Square, London, WC1B 3DP, UK
29 Earlsfort Terrace, Dublin 2, Ireland

BLOOMSBURY, BLOOMSBURY LIBRARIES UNLIMITED and the Diana logo are trademarks of Bloomsbury Publishing Plc.

First published in the United States of America 2025

Library of Congress Cataloging-in-Publication Data
Names: Fernandez, Michael (Librarian), author. | Serrano, Amauri, author.
Title: Streaming video collection development and management / Michael Fernandez, Amauri Serrano.
Description: New York : Bloomsbury Libraries Unlimited, 2025. | Includes bibliographical references and index.
Identifiers: LCCN 2024029787 (print) | LCCN 2024029788 (ebook) | ISBN 9781440880858 (paperback) | ISBN 9798216172390 (epub) | ISBN 9781440880865 (pdf)
Subjects: LCSH: Libraries--United States--Special collections--Streaming video. | Academic libraries--Collection development--United States. |
BISAC: LANGUAGE ARTS & DISCIPLINES / Library & Information Science / Collection Development | TECHNOLOGY & ENGINEERING / Television & Video
Classification: LCC Z692.S77 F47 2025 (print) | LCC Z692.S77 (ebook) | DDC 025.17/4--dc23/eng/20240819
LC record available at https://lccn.loc.gov/2024029787
LC ebook record available at https://lccn.loc.gov/2024029788

Typeset by Deanta Global Publishing Services, Chennai, India
Printed and bound in the United States of America

To find out more about our authors and books, visit www.bloomsbury.com and sign up for our newsletters.

Contents

Acknowledgments

This book would not have been possible if it were not for the contributions of many of our colleagues who lent their efforts during the process of writing and revising the manuscript. The authors would like to use this opportunity to thank those who generously provided their knowledge, experience, and most importantly, their time to this project.

Many of the supplementary models and templates included in the appendices were produced by our Yale University Library colleagues. The authors would like to thank Jennifer Bazeley, Sandra Aya Enimil, and Rowena Griem for kindly permitting them to be adapted as practical supplements that readers can apply within their own libraries.

It was highly important for the authors to go beyond their own experiences and seek out perspectives from colleagues working at other libraries to better encompass the state of practice for managing streaming video collections. Many of these participants were solicited through conferences or listservs, and kindly responded to our cold calls and interview requests. We are grateful for the collegial sharing of knowledge from the following: Theresa Arndt, Whitney Bevill, Mattie Cook-Hildebrandt, Jennifer Cox, Summer Durrant, Kristina Edwards, Ed Garcia, Genya O'Gara, Benjamin Paul, Andrea Rodgers, Michael Rodriguez, Lisa Sallee, Scott Stangroom, Gabriel Stetson, Timothy Streasick, Jane Strudwick, Merlita Taitague, Shannon Tennant, and John Wisell. The authors thoroughly enjoyed talking about streaming video with all of our interviewees and deeply appreciate being part of a professional community that values the exchange of ideas and experience.

Introduction

Streaming video is an increasingly inescapable part of our lives. We are constantly streaming content, whether on our smartphones, laptops, or home entertainment consoles. We stream content at all times of the day, on the subway, in the office, and in our living rooms. At the end of 2021, it was estimated that 109.4 million US households, or 85 percent, subscribed to at least one video-on-demand service (Sangari 2022). The options for streaming video seem endless and omnipresent, populating our social media feeds and bolstering an ever-expanding array of content platforms. Streaming video is increasingly the default medium for communication, for entertainment, and, perhaps unsurprisingly, for education.

Streaming video is not new to the academic environment but recent years have seen an exponential growth in the number of platforms and titles available for streaming. For libraries, this has meant an increasingly complex acquisitions landscape, with more content providers occupying the marketplace and larger portions of the budget dedicated to streaming. Given the ubiquity of streaming in everyday life, it is no surprise that users increasingly expect video content to be available online and on demand. Streaming video is increasingly integrated into coursework across all subject and discipline areas. As a result, libraries are faced with new challenges in building and managing streaming video collections.

While there was no singular tipping point for streaming video in libraries, the onset of the COVID-19 global pandemic in the early months of 2020 functioned as an accelerant for the already burgeoning format. Libraries found themselves quickly pivoting to the virtual space, often needing to source online equivalents of physical media collections rendered inaccessible for the foreseeable future. As in-person activity and instruction gradually resumed, the immediate need for streaming video tapered off, somewhat, but user demand still remains at a high level (Foley 2023).

The intent of this book is to address the challenges of managing streaming video content in a straightforward and practical manner. The authors have experienced the growth of streaming video in libraries firsthand, working in public and private academic libraries of varying sizes, budgets, and research focuses. In our respective roles at Yale University Library, we handled all aspects of streaming video collections, from initial request through delivery of access to content, as well as ongoing management. As streaming video management became increasingly core to our job roles, we noted an overall lack of training materials and were often creating new documentation and workflows from scratch that were unique to the format. Like many libraries, Yale does not have a dedicated Media Librarian, placing the responsibility for the management of streaming video squarely within the areas of collection development and electronic resources. The need for a book like this was readily evident to us, and we similarly hope it will serve to aid any library that is acquiring streaming video.

As streaming video management increasingly becomes part of workflows for library staff, the authors sought to document best practices for structuring those workflows and offer guidance in onboarding and training staff to meet the demands of heightened streaming acquisition. From the initial emergence of streaming video collections in libraries, it was quickly recognized that a collaborative approach was needed, as the format straddled areas of responsibility including collection development, technical services, and systems (Schroeder and Williamsen 2011). An early survey in 2013 found that primary responsibility for streaming video was equally or more likely to be delegated to a collection development or acquisitions librarian than it was to a media librarian, particularly with regard to licensing content (farrelly and Hutchison 2014, 73). A 2022 survey of academic libraries found that the most common job title bearing responsibility for licensing and acquisition of streaming video was "Electronic Resources Librarian," indicated by 39.15 percent of respondents. Other common job titles responsible for streaming video included "Acquisitions Librarian" and "Collection Development Librarian"; additionally, non-librarian staff within acquisitions, technical services, and e-resources units were also listed by respondents, supporting the collaborative and team-based approach frequently applied to the streaming format (Austin and Thornton 2022, 248–50). Recognizing that our areas of electronic resources and collection development are often tasked with streaming video management, we applied that perspective to make this book a valuable resource for workers within those areas.

Audience

The intended readers for this book are library practitioners who work with collection development or acquisition of streaming video. Those workers whose job duties entail selecting, licensing, acquiring, and managing streaming content will find useful guidelines for maximizing library efficiency in these areas. Library workers tend to wear many hats, particularly at smaller libraries, so they may already have selection and acquisition duties, or they may find streaming video suddenly added to their portfolio. For library staff just coming into these areas, this book will be an engaging primer and how-to manual to get them started with building and managing a streaming video collection. For those staff members already experienced with streaming video, this book can be a useful reference source to build upon their knowledge and incorporate best practices for collection development and acquisition of streaming into their existing workflows.

Drawing from the authors' experiences working in college and university libraries, many of the examples in the book are focused on institutions of higher learning, be they community colleges, research institutions, or specialized schools. While that is the primary perspective, much of the ground covered will have applicability to public and school libraries as well. In addition to their own professional experience, the authors drew heavily from the published literature, conference presentations, and, crucially, focused interviews with colleagues at other libraries. We felt it was important to seek out diverse perspectives on how streaming video is managed at different types of libraries, including university libraries of varying sizes and collection budgets, as well as community colleges, public libraries, and library consortia. We sought to reflect these perspectives throughout in our writing.

Additionally, we have spotlighted these institutional studies of streaming video practice in our "Streaming Vignettes," which are interspersed between chapters. Each vignette focuses on an individual library, providing a local snapshot of what streaming content is being acquired, how the collection is managed currently, and where future challenges may lie ahead. While budgets and collection focus obviously vary widely, we found that many of the common challenges for streaming video collections were shared across all different library types. Based on these shared experiences, we feel that the fundamental aspects of streaming video management covered by this book will resonate with readers and that the practical guidance on how to handle the challenges of this format will be replicable, regardless of library size and budget.

Organization of the Book

In organizing the chapters of this book, the authors sought to replicate the sequence of activities from request to delivery of streaming video content. The sequencing was also heavily informed by the electronic resources life cycle, as originally envisioned by Pesch (Pesch 2008), and built upon through successive iterations of Emery, Stone, and McCracken's TERMS project (Emery and Stone 2013; Emery, Stone, and McCracken 2020). The initial chapters cover the investigation stage of the life cycle. Chapter 1 discusses the evaluation of content, Chapter 2 outlines the numerous different acquisition models that can be pursued, and Chapter 3 addresses the universal concern of managing budgets. Pivoting to the acquisition stages of the life cycle, Chapter 4 covers the all-important process of licensing streaming video, which will inform all aspects of how the content can (and cannot) be used. Chapter 5 is a workflow-centered examination of how streaming content is ordered and received, while Chapter 6 addresses many of the specialized facets of streaming video content hosted/managed locally by the library. Moving onto the implementation stage, Chapter 7 discusses both providing access through library discovery tools and maintaining that access through troubleshooting. Related to access, accessibility is a critically important area of streaming video management, and so we have devoted the entirety of Chapter 8 to examining the best practices for making streaming content accessible to all users. Like any other electronic resource format, periodic assessment is necessary for streaming video—Chapter 9 covers how to perform this assessment with both quantitative and qualitative metrics. Finally, Chapter 10 looks at the future of streaming video in libraries and the challenges posed by the format to core principles of librarianship: scholarly sharing, openness, affordability of educational materials, and preservation.

Reading the book cover to cover will provide an overview of the successive stages involved in managing streaming video collections. With that said, the authors recognize that library workers may only deal directly with certain stages of the process themselves; the sequencing also supports reading a chapter or two at a time. Due to this, some repetition of concepts naturally occurs across chapters. Where it is applicable, we direct readers to corresponding chapters for more robust coverage of certain topics; there is also a glossary of frequently used terms for reference. Additionally, we wanted to provide readers with practical tools, and so the appendices contain selection guidelines, a model license, and a cataloging template. These have all been adapted from documentation used at the authors' institution and can be applied by readers at their own libraries.

Notes on Scope and Terminology

Our perspective as an academic library based in the United States informs the scope of this work. This is most notably evident in the area of copyright; there are frequent mentions of US copyright law, particularly within Chapter 4 on licensing. Our overall guidance on the application of fair use for streaming—with the customary disclaimer that we are not lawyers—is based on a conservative interpretation of Section 107 of the US Copyright Act, as supported by the majority of the existing literature on licensing. However, we acknowledge that readers outside the United States will be subject to different laws surrounding copyright and intellectual property rights. For example, Canada's fair dealing rights for educational use incorporate an entirely different set of determining factors than US fair use ("Fair Dealing in Canada" n.d.). References to accessibility laws in the United States such as the Americans with Disabilities Act (ADA) and Section 508 of the Rehabilitation Act are made in Chapter 8, although much of the guidance for best practice is taken from the more robust Web Content Accessibility Guidelines (WCAG) 2.1, as developed by the international web standards organizations, the World Wide Web Consortium (W3C).

Libraries license streaming video across a range of subject areas from a variety of content providers. Much of the most in-demand streaming videos is feature film and documentary content, particularly for public libraries. The most commonly used streaming providers offer multidisciplinary content. A 2021 survey of academic libraries found the two providers most utilized by libraries were Kanopy (83 percent of respondents) and Alexander Street (80 percent of respondents). Kanopy and Alexander Street are both multidisciplinary platforms with a large amount of feature film and documentary content. Many of the other popular providers indicated by the survey—Films on Demand, Swank, and Docuseek—also offer feature and documentary content that is multidisciplinary in scope (Cooper, Ruediger, and Skinner 2022, 6). More specialized streaming providers in STEM (Journal of Visualized Experiments [JoVE], Psychotherapy.net) and the humanities (Digital Theatre+, Met Opera on Demand) tend to get picked up by libraries that have more subject-focused collections. Our approach to streaming content in this book is generalized and subject agnostic, with an awareness that feature and documentary content is the most heavily acquired by libraries.

In library parlance, "vendor" is the common term for providers that sell materials or content to libraries. Streaming video can be licensed from sellers of myriad sizes, from large entities like Kanopy (owned by OverDrive) and Alexander Street (owned by ProQuest/Clarivate) to small independent distributors or filmmakers who directly sell their own films that are the most likely candidates for local hosting by libraries. The authors tend to favor the term "vendor" throughout, but also use the terms "content provider" and "distributor" interchangeably, based on how the phrasing contributes to readability or situational context. For example, "distributor" is generally preferred over "vendor" when describing local hosting of content, as such content is frequently licensed from smaller entities, and so "distributor" carries a more appropriate connotation. While the glossary section outlines specific role-based distinctions among these terms, multiple roles often apply to the same entity. Women Make Movies is an example of a film distributor that sometimes acts as a vendor, selling content directly to customers, but also distributes through third-party vendors, making portions of their catalog available for sale through Kanopy.

References

Austin, Cathy and Bonnie Thornton. 2022. "Streaming Video and Academic Libraries: A Survey on Commonalities in Acquisitions." *Journal of Electronic Resources Librarianship* 34 (3): 245–55. https://doi.org/10.1080/1941126X.2022.2099040.

Cooper, Danielle, Dylan Ruediger, and Makala Skinner. 2022. "Streaming Media Licensing and Purchasing Practices at Academic Libraries: Survey Results." *Ithaka S+R*. https://doi.org/10.18665/sr.316793.

Emery, Jill and Graham Stone. 2013. "TERMS: Techniques for Electronic Resources Management." *Library Technology Reports* 49 (2): 5–43. https://doi.org/10.5860/ltr.49n2.

Emery, Jill, Graham Stone, and Peter McCracken. 2020. *Techniques for Electronic Resource Management: TERMS and the Transition to Open*. Chicago: ALA Editions.

"Fair Dealing in Canada: Myths & Facts." n.d. Canadian Association of Research Libraries. Accessed September 28, 2023. https://cdn.dal.ca/content/dam/dalhousie/pdf/library

/services/copyright-office/CARL_FD_myths_facts_EN.PDF.lt_aa43df8109ad0acf309d72a1e47c949c.res/CARL_FD_myths_facts_EN.PDF.

farrelly, deg and Jane Hutchison. 2014. "Academic Library Streaming Video: Key Findings from the National Survey." *Against the Grain* 26 (5): 73–5. https://doi.org/10.7771/2380-176X.6852.

Foley, Marianne. 2023. "Evaluation of Streaming Video Usage in a University Library Before, During, and After the COVID-19 Pandemic Lockdown." *The Journal of Academic Librarianship* 49 (3): 1–6. https://doi.org/10.1016/j.acalib.2023.102698.

Pesch, Oliver. 2008. "Library Standards and E-Resource Management: A Survey of Current Initiatives and Standards Efforts." *The Serials Librarian* 55 (3): 481–6. https://doi.org/10.1080/03615260802059965.

Sangari, Nicole. 2022. "85% of US Households have a Video Subscription Service." January 26, 2022. Kantar. https://www.kantar.com/north-america/inspiration/technology/85-per-cent-of-us-households-have-a-video-subscription-service.

Schroeder, Rebecca and Julie Williamsen. 2011. "Streaming Video: The Collaborative Convergence of Technical Services, Collection Development, and Information Technology in the Academic Library." *Collection Management* 36 (2): 89–106. https://doi.org/10.1080/01462679.2011.554128.

1 Collection Strategies

Introduction

Libraries have been building and providing access to video collections in a variety of formats for decades. These collections have been, and continue to be, shaped by local needs and interests, evolving technologies, and changing business models. Video-based learning has been a staple in higher education for many years. Over the past decade, as universities have expanded distance education and online instruction, library media collections have increasingly shifted toward streaming video (Brancolini 2002; Carmichael, Reid, and Karpicke 2018). Survey data indicate that the number of academic libraries offering streaming video tripled from 33 percent in 2010 to 96.7 percent in 2021 (farrelly and Hutchison 2014; Tanasse 2021). Similarly, data from the annual *Library Journal* public libraries materials survey show that public library spending on streaming media doubled between 2019 and 2021 (Wyatt 2022).

Streaming video provides several benefits including ease of access, pedagogical advantages, and improved accessibility for those who are deaf, have vision disabilities, and motor impairments (e.g., captions, audio descriptions, transcripts, and keyboard controls) (Spina 2021). It can also address and mitigate the obsolescence of video playback equipment or the removal of playback equipment in classrooms (farrelly 2014); however, streaming video also presents a series of challenges for libraries. The biggest challenge for both academic and public libraries is balancing user demand with the high cost to both license the content and the indirect staffing and infrastructure costs to source, order, and provide access (Tanasse 2021; Cooper, Ruediger, and Skinner 2022; Wyatt 2022). The number of ever-changing acquisition models, variety of pricing models, and lack of licensing options have made it difficult for libraries to source, provide access, and budget for video. Most importantly, the rapid transition from physical to streaming media has transformed library collecting strategies and practices from a model focused on ownership, curation, and preservation of media to what Handman calls a "service-based model of access to licensed resources," which he argues "entail(s) a major rethinking . . . of the budgeting process, the nature of services and staffing, and many of the [library's] traditional roles" (Handman 2010, 326).

This chapter discusses and explores the reasons why libraries license streaming video and how community needs and collection development philosophies shape collecting strategies. It provides an overview of the educational streaming video market and best practices for sourcing streaming media from content providers. Finally, it offers practical guidance on drafting a video collection development policy and selection guidelines to ensure consistency of practice and alignment with institutional collection development goals and objectives.

Purpose of Streaming Video in Libraries

Libraries license streaming video for a variety of reasons including curricular and research support, public programming, entertainment, and cultural enrichment. Given competing needs and limited budgets and staffing, libraries must decide what video content and formats (i.e., streaming or physical) best meet the diverse needs of their user communities. These needs are then translated into specific collection development goals and priorities, and as institutional needs change, so should collecting strategies.

Curricular Support

As evidenced in the literature and supported by survey data, academic libraries primarily license streaming video for curricular use (Levenson and Lombardo 2023; Tanasse 2021; Cooper, Ruediger, and Skinner 2022). Streaming video has become a vital pedagogical tool in higher education due to several factors: the growth of distance education, innovative approaches to online teaching and learning, the convenience of 24/7 access for multiple users, and a generation of students accustomed to streaming for both personal and academic purposes. Studies have shown that faculty and students recognize the pedagogical benefits of video, which are further enhanced by digital access (Carmichael, Reid, and Karpicke 2018; Valenti, Tennille, and Mandernach 2019; MacDougall and Ruediger 2023). For faculty, streaming video supports engagement with students, accommodates diverse learning styles, and aids in teaching discipline-specific skills. For students, it allows for self-paced learning, reinforces key concepts, and provides diverse perspectives on various topics (MacDougall and Ruediger 2023). New instructional methods, such as the flipped classroom, blended learning, and hybrid courses, have significantly influenced the use of streaming video in teaching. The flipped classroom reverses the traditional model: students learn foundational concepts outside of class, freeing class time for discussion and application. Blended courses integrate multimedia resources with traditional classroom instruction (Carmichael, Reid, and Karpicke 2018, 7), while hybrid courses mix online and in-person sessions. These approaches reflect how streaming video has transformed both teaching practices and student learning experiences.

Research Support

The nature of some disciplines requires video for research and teaching such as the performing arts, film studies, and video art. Video is also important for observational studies in disciplines that study human behavior like ethnography and psychology or in disciplines that find value in visual examples of experiments like in the life and physical sciences. Discipline-specific content is often found in streaming video packages or available in discipline-specific collections and platforms such as Alexander Street's Ethnographic Video Online, Psychotherapy.net, or the Journal of Visualized Experiments (JoVE). In addition to collections and platforms, researchers may be interested in individual films that have been recently released (e.g., topic-specific—LGBT rights), older content that Shirts describes as meant "to get a glimpse at the past" (e.g., theatrical performances, period newsreels), or unique and foreign titles that are not easily sourced or held by a limited number of institutions (Shirts 2019, 3). For certain disciplines such as film and media studies, video is a core resource where specialized titles, alternate versions, or special editions are regularly requested such as a director's cut or a version with supplementary content (e.g., interviews with the actors). Because of their specialized nature, many titles requested for research purposes will likely get little use, which is a budgetary concern for titles that are only available streaming and are expensive (Shirts 2019). Libraries may also end up duplicating content by purchasing DVDs of titles they already have access to streaming, in order to fill requests for special editions.

Public Programming

Campus groups often screen films for educational purposes, public engagement and awareness of a topic, or simply for entertainment purposes; however, they are often unaware of copyright restrictions or the need to secure public performance rights (PPR) for film screenings made open to the public. Although copyright and PPR are discussed at greater length in Chapter 4, a brief overview of allowed use under US copyright law is useful to illustrate how intended use impacts collection development decisions. US copyright law allows for face-to-face instruction or as Wahl describes, "the performance (i.e., showing) of films in the classroom without securing permission from the rights holder as long as conditions such as showing only to students enrolled in the course and using legally obtained copies are in place" (Wahl 2019, 79). The Technology, Education and Copyright Harmonization Act ("TEACH") complements face-to-face instruction and provides an exception for accredited non-profit US educational institutions to make certain copyrighted materials available for online instruction or distance learning purposes. However, as Willi Hooper notes PPR are needed if "a viewing occurs: (1) at any place open to the public or (2) where a substantial number of persons outside of a normal circle of family and its social acquaintances is gathered" (Willi Hooper 2018, 4). Although some licensed streaming content may include PPR (e.g., select Kanopy films or documentaries available from Docuseek), PPR can also be purchased if available. Some libraries may have a "no PPR purchase policy" while others may purchase PPR based on expected use, available funding, or if available at a reasonable price. In an effort to facilitate access and support for campus screenings, some libraries show the availability of PPR information in their catalog records.

Infotainment

Entertainment is often one of the primary reasons that public libraries license streaming video, but media content in public libraries also serves the informational and cultural needs of their communities, as is often reflected in their mission statements that emphasize learning and community building (Fraser-Arnott 2021) and in their streaming services such as Kanopy and Hoopla that offer an array of content for different populations (e.g., children, teens, and adults). It is not surprising that considering the high cost of streaming, academic libraries have chosen to prioritize licensing streaming video for teaching and research purposes. However, a major appeal of subscribed aggregated streaming collections (e.g., Films on Demand and Academic Video Online) is their versatility: they support educational needs, provide content for personal enrichment and entertainment, and serve as a user engagement tool.

Collection Development Philosophy and Strategy

A library's collection development philosophy refers to the principles and guidelines a library uses to select, acquire, and manage library collections. It guides decision-making and aligns with the library's mission which "defines the library's purpose and describes who the library serves" (American Library Association 2017). It guides the library's collection development efforts and defines the scope of the collection and collecting goals and priorities. For example, a collection development philosophy for a large research library may focus on building and preserving collections, whereas a community college library may focus on maintaining a curriculum-focused collection.

The philosophy informs the library's collecting strategy and approaches to building and providing access to information resources. These approaches are often referred to as "just-in-case" and "just-in-time," and Gary Handman, former Director of the Media Resource Center at UC Berkeley, adds a third approach: "a circulating collection," which emphasizes the currency and popularity of materials, selected to meet the broad entertainment and educational needs of community users (Handman 2010, 326). These general approaches inform decisions regarding preferred acquisition models and the preference to own or lease content. A just-in-case approach is a traditional collecting approach that is often found in large research libraries that aims to build broad collections to meet the perceived needs of users as well as preserve the cultural record. Selection is based on what the library deems valuable to its users and the broader scholarly community. As Handman notes, a just-in-case approach requires a large financial commitment and the staff expertise and time to select materials. On the other hand, a just-in-time approach aims to provide access to materials at a point of need. Libraries may take a just-in-time approach for budgetary reasons (fluctuating budget, high cost of materials, steep inflationary increases, and effective stewardship of collections budget), limited staffing capacity and subject expertise, and the desire to be more responsive to user needs.

A library's approach to collection development will inform which acquisition models they use to license streaming video, and, in some cases, the decision to acquire physical media for preservation purposes. Libraries often take a combination of approaches for different purposes. For example, a library may take a just-in-time approach for course reserves, a just-in-case approach for content essential to teaching and learning in a discipline (e.g., theater), and a "circulating approach" when subscribing to a collection that supports the curriculum while also providing entertainment value (e.g., Swank's Top 1000 collection, a collection of major studio feature films). Although collecting approaches are useful in shaping the library's collection development strategy, libraries may have little choice in selecting an acquisition model for requested films often due to a lack of options.

As noted earlier, streaming offers numerous benefits for teaching, learning, and research. However, for all its benefits, an all-in streaming approach (i.e., if streaming is not available, the library will not purchase a physical copy) to video collections has its drawbacks in the short- and long-term (Shirts 2019). While user preference is a key factor in collection decisions, it also needs to be weighed against the reality of streaming providers losing distribution rights and the need for long-term preservation of video. Although the number of streaming titles available for perpetual purchase has grown, most streaming titles are only available as term licenses or as part of subscriptions (i.e., streaming delivered as a service). For those libraries that aim to build film collections for long-term use, current streaming options make it difficult to collect and preserve films deemed valuable.

For select titles, DVDs may remain the only viable option. Proctor recommends that libraries not only consider usage to make collection development decisions but also the "enduring value of the film" (Proctor 2018, 84). She further states, "libraries need to be thoughtful about the films that they can envision being in demand five, ten, or twenty years from now because if libraries stop buying DVDs, they may find film collections severely lacking when its too late to go back and purchase them" (89). Another factor to consider with an all-in streaming approach is the lack of foreign and independent films on provider platforms, which has implications for teaching, research, and learning. As Brian Hu, a professor at San Diego State University, highlights, film studies relies on "film examples to illustrate the work of institutions, artists, and industries, which are not easily substitutable," and notes that he often uses DVDs because there are no streaming options available for films he uses in his courses (Hu 2022). In addition, libraries that support film studies programs often acquire DVDs because they have better picture quality, provide different versions of a work (e.g., director's cut or extended versions), and often include extras such as interviews and deleted scenes. Therefore, in order to achieve their collection development goals and priorities, some libraries must take a hybrid approach to collecting video, one that balances streaming with physical video acquisitions (Shirts 2019).

Content Providers, Acquisition Models, and Sourcing

Content Providers

The streaming vendor landscape is diverse and fluid, with numerous content providers of different sizes ranging from large academic vendors that license thousands of titles to small independent distributors that specialize in a particular area (e.g., region of world, media art, etc.) or genre of film. Some offer discipline- and subject-specific content such as JoVE, a scientific video magazine, or Met Opera on Demand, a streaming collection of Met performances. Although there are many streaming media providers, four dominate the content market—Kanopy, Alexander Street, Films on Demand, and Swank (Dixon 2017; Cooper, Ruediger, and Skinner 2022). These providers have their own platforms, offer a broad array of content, a variety of acquisition models, and some provide hosting services. Independent distributors primarily offer exclusive content in niche areas such as social justice documentaries or Arab and Middle Eastern films. Most of their content is available for perpetual purchase but requires locally hosting. A unique aspect of streaming video is that libraries regularly license content directly from filmmakers or rights holders by faculty request. In addition to traditional selection criteria such as content, price, and relevance to research and the curriculum, acquisition models play a role in how and why streaming content is licensed. A select list of educational streaming providers and acquisition models offered can be found in Appendix D.

Although the amount of streaming video content available for institutional use continues to grow, so do the number of direct-to-consumer platforms (e.g., Netflix, Disney+, Hulu) as well as the amount of original programming they produce. These companies invest heavily in producing exclusive content to attract and retain subscribers. Content on these platforms is unstable, and titles are regularly added and removed, including exclusive titles. Unfortunately, the vast majority of these companies do not license their content to libraries (as of this publication), which not only has implications for long-term preservation but also affects the library's ability to fill course reserve requests (Joint Digital Content Working Group 2020). For example, Rodgers notes that educational streaming providers such as Kanopy and Alexander Street do not offer the variety of feature films required by film studies programs, but direct-to-consumer streaming platforms that provide this content do not license to libraries (Rodgers 2019, 582). A 2017 *Library Journal* survey found that nearly half of faculty assign videos on direct-to-consumer platforms in their courses (Dixon 2017). In fact, faculty have been known to screen content on commercial streaming platforms in class, which is a potential violation of copyright (Ezor 2013). Currently, Netflix is the only commercial streaming service to offer a select number of documentaries for educational use (Netflix n.d.). Some libraries reference direct-to-consumer platforms in their collection development policies or online streaming video guides. They often highlight that class screenings of content on commercial streaming platforms are a violation of subscriber terms of use, and note that the library can purchase DVDs for content on direct-to-consumer platforms and recommend that users pay for personal access when DVDs are not available (King 2024; McCleskey 2022; University of Michigan-Flint 2024). JustWatch (https://www.justwatch.com/) and Reelgood (https://reelgood.com/) are streaming service search engines that are useful to identify films available on commercial streaming platforms that are not available for institutional licensing.

Streaming Video Acquisition Models

In addition to traditional selection criteria such as content, price, and relevance to research and the curriculum, acquisition models play a central role in how and why streaming content is licensed (Stone and Emery 2013). Libraries may choose to use multiple models based on the use case—for example, term licenses for course reserves and perpetual access for faculty research requests that directly support the curriculum. Although acquisition models are discussed in Chapter 2, a general overview is provided here as they are a core component of a library's video collection development strategy.

In some respects, streaming video is similar to the acquisition of e-books and e-journals. It is expensive, can be licensed title-by-title or as part of packages, can be leased or owned, and may not be available for institutional purchase in electronic format (e.g., e-textbooks). However, purchase models for streaming video differ in a few key ways; they often require more staff time and expertise to source the content, negotiate licenses, and provide access. Acquisition models fall into five broad categories: perpetual, short-term, subscriptions, use-driven, and unavailable. Although "unavailable" is not an acquisition model per se, it captures those titles that are not available streaming for institutional use (e.g., Netflix Originals) or are not available for streaming at all (Ballestro 2019). Unfortunately, a significant number of films fall under the "unavailable" category, requiring libraries to purchase physical formats to meet user needs. A well-rounded collection development strategy will use multiple acquisition models to achieve desired goals—collection building, access to course-related materials, research support, public programming, and user engagement and entertainment.

Perpetual

A limited number of films are available for perpetual purchase either as single titles or collections. They are often documentaries, films from independent distributors, and those sourced directly from filmmakers. However, large streaming providers like Alexander Street offer subject and thematic collections for perpetual purchase. Feature films from major studios are rarely available for perpetual purchase, but some international films from independent distributors are available for perpetual purchase either title-by-title or as collections. Perpetually purchased streaming video collections are usually hosted on provider platforms and may require an annual hosting fee. Single titles, on the other hand, can be purchased and accessed on provider platforms (i.e., perpetual access license—PAL) or managed locally (i.e., digital site license—DSL). Like collections, providers may charge an annual hosting fee for individual films purchased hosted on their platform or a flat platform fee. There are no industry pricing standards, and the cost to perpetually license individual films can vary considerably. The cost of perpetual licenses can vary based on the amount of content included in collections and packages, the type of content (e.g., film featured at an international film festival), and the distributor. The average cost for a single PAL can range from $300 to $500 per title. The range in pricing for perpetual DSL can vary considerably and can range on average from $100 to $1000.

Short-Term

Short-term licenses are available for single titles and multi-part series that are licensed for a defined period, usually one to three years, and to a lesser extent five; however, the license period can be as short as a one-time screening. Videos are usually hosted on a provider platform, but a small number of independent distributors offer short-term licenses for videos that require local hosting. A short-term license is often used for course reserves and public programming or screenings. On average, short-term licenses range from $150 to $250.

Subscriptions

Subscriptions are the most common acquisition model found in academic libraries (Tanasse 2021; Cooper, Ruediger, and Skinner 2022). Grombly defines a subscription as "a renewable annual or continuing contract providing access to a specific list of titles and/or collections" (Grombly 2020, 164). However, some subscriptions act like aggregated databases, in which content is regularly added and removed. Subscriptions provide unlimited access to a large amount of content at a fixed price. Vendors often provide MARC records for subscribed collections, reducing staff time and facilitating discovery. Streaming video subscription pricing models are similar to other e-resource database subscriptions, which are often based on the number of full-time equivalent (FTE) students or tiered pricing by Carnegie class. They range in price depending on the amount of content and can cost over $25,000.

Use-Driven

Use-driven acquisitions (UDA) is an umbrella term used to describe models in which videos are licensed based on user discovery rather than the anticipation of use (Carrico, Leonard, and Gallagher 2016, xiii). Streaming video UDA models include patron-driven acquisitions (PDA), evidence-based acquisitions (EBA), and pay-per-use (PPU). In a PDA model, sometimes referred to as demand-driven acquisitions (DDA), users have access to a broad array of content that can range from the provider's entire catalog to a customized list of titles (e.g., Kanopy's Smart PDA). In this model, a short-term license is triggered for a period of one to three years based on the number and length of views, with no library mediation. However, to control costs, some libraries have opted for a mediated PDA model, in which library staff review and approve requests submitted on the platform. In an EBA model, the library pays an upfront fee and users have access to the provider's entire catalog or a pre-selected collection of titles for a limited period (e.g., four to twelve months). At the end of the period, the library selects videos for perpetual purchase equal to the amount of the upfront fee. Although selection decisions for EBA programs are often based on usage, libraries may use other selection criteria, such as subject matter or distributor. Public libraries often use a PPU model, in which the library pays a fee, often ranging from $2 to $6, each time a video is watched. Libraries can often set the viewing period (e.g., three days) and the number of films users can watch during a defined period (e.g., five films per month).

Sourcing

Licensing streaming for course reserves is a top priority for academic libraries. Many films used for teaching are available on institutional streaming platforms and from known educational distributors. Unfortunately, streaming providers often lose distribution rights and remove titles from subscribed collections, including films used in courses, requiring library staff to scramble to find alternative streaming access. Ideally, with a bit of searching, the film will be available from another provider, but the sourcing process is not always easy. Faculty exposure and access to streaming video via commercial streaming services, screenings at international film festivals, while conducting research abroad, or simply via a trailer on YouTube have made the work of sourcing video and managing faculty expectations a true challenge. Educational streaming providers do not provide all films needed for teaching and research; this is especially true for rare, foreign, or independent films. Finding the streaming version of a film can be simple, but it can also be time-consuming and require a team approach to locate the rights holder and negotiate pricing and terms of use. For example, a request for a Polish television series may require librarians with foreign language skills, knowledge of video acquisitions, and licensing expertise to work together to locate the rights holders, negotiate pricing and terms of use, and discuss the technical aspects of file delivery and caption files. Below are some best practices for sourcing streaming video.

Step 1: Determine if the library currently has access

1. Search the library catalog to identify if the film has already been licensed.
2. If the film is requested for a course, ask the instructor how long they will use the film. If the film is needed for multiple semesters, then a three-year or perpetual license may be a good option. If the film has already been licensed by the library, this is an opportunity to confirm that the license dates cover the entire period needed. Some content providers display expiration dates on their platforms, which helps instructors plan and reduces the probability that students will lose access during the semester.
3. Before starting the search, consider the subject, category of film (e.g., feature film, documentary, or counseling session), and distributor to more easily locate the film in library licensed collections and/or platforms (e.g., Kanopy for a Kino Lorber film and Docuseek for a Pragda film). It is useful to become familiar with suppliers/producers commonly found on provider platforms to locate films easily and quickly. For example, Swank distributes films from Columbia Pictures, Alexander Street-Sony Picture Classics, Kanopy—the Criterion Collection, and Criterion Pictures—20th Century Fox.
4. Search library streaming platforms (e.g., Kanopy, Alexander Street, Films on Demand, Swank). Films are regularly added to subscribed collections, but there may be a lag in uploading the catalog records. Some libraries do not upload records for certain content providers or certain platforms. For example, some collections lack stable URLs for individual videos and are cataloged as entire collections, rather than cataloging each video separately (e.g., Electronic Arts Intermix). Some libraries also choose not to add records for PDA plans to limit the number of licenses triggered, control costs, as well as reduce the amount of staff time to maintain accurate records (Lowe et al. 2021, 121; Prelitz 2022, 325).

Step 2: Source films not available in the library's collection

1. Search WorldCat to determine if another library has streaming access in order to identify the streaming provider. This is also an opportunity to gather bibliographic information (e.g., producer, distributor, or release date), which will help locate the rights holder.
2. Ask colleagues. The Film and Media Roundtable VideoLib list (https://guides.lib.berkeley.edu/FMRTlists) is an excellent resource for librarians seeking advice on video-related topics including sourcing streaming access to specific films. The list is heavily used, quite effective, and can save a considerable amount of time.

3. Gather information online. Search IMDb (https://www.imdb.com/) to gather information about the requested film (e.g., synopsis, director, release date, etc.). Search streaming service search engines—JustWatch (https://www.justwatch.com/) and Reelgood (https://reelgood.com/) to locate the film on a consumer-facing platform. Search the internet with information gathered. Depending on the type of film, social media may also be an option (e.g., contemporary dance).
4. If you are able to find contact information, contact the filmmaker or distributor, ask if the film is available on a streaming platform accessible to educational institutions. If this is not the case, ask if they are the copyright holder and can grant streaming access.

Sample Streaming Request

I work at [insert university name] OR I am the [insert title and university] and the library is trying to obtain streaming rights [insert title of film, director]. A faculty member in [discipline] has requested streaming access for teaching and research purposes.

We would like to know if the film is available on a streaming platform that provides access to educational institutions or if we can host the film on a password-protected server and make it available to university-authorized users. If you are not the copyright holder, can you direct me to the appropriate person or distributor?

Collection Development Policy and Guidelines

Gregory defines collection development policies "as a blueprint for the operations of the library as whole, for it is through these policies that the library carries out its central tasks of acquiring, organizing, and managing library materials" (Gregory 2019, 29). CD policies provide the framework for defining a library's short- and long-term collection goals for new acquisitions and the maintenance of existing materials. They serve multiple functions: guide decision-making, help coordinate selection, provide consistency of approach, assist in allocating funds, demonstrate how the budget is spent, serve as a training and communication tool for library staff, and help manage user expectations. CD policies also describe how the collection supports the library's mission; state collection development goals and objectives; define the scope of the collection; highlight collection strengths; and outline priorities for the selection, acquisition, and evaluation of resources (Shaw 2011, 166). These policies can be broad, comprehensive, or subject-, format-, or collection-specific (e.g., oral history collection).

The ACRL *Guidelines for Media Resources in Academic Libraries* recommends that libraries have a written collection development policy statement to guide the selection and acquisition of media resources that outlines media-specific priorities and guidelines (ACRL 2018). A media-specific policy addresses the unique aspects of streaming video selection and management that are relevant to the library and its user community. The word policy is used here to denote documented CD guidelines which may take the form of a standalone video/streaming video policy that references criteria in subject policies. The policy may be embedded in a streaming video LibGuide that also includes lists of licensed streaming video collections and provides general information regarding the use of streaming video. Streaming video selection criteria could be included in a library's general CD policy (i.e., the entire collection). The structure and amount of detail in the policy will differ by library based on the size and breadth of the collection, the number of staff involved in the selection and management of streaming video, and the library's collection development philosophy (e.g., collection building and/or access at the point of need).

The policy should have a clear statement of purpose, be practical, as well as easy to update to reflect changes in technology, acquisition models, staffing, and user needs. It should provide consistency of practice, serve as a guide to what is licensed and what is not licensed, outline selection responsibilities, and provide criteria, considerations, and preferences for selection. For those libraries that support film and media studies programs, the policy should address video as both a format (i.e., content) and film as a subject (e.g., film history, theory, and aesthetics) which may inform the choice of format for particular use cases. Considering the high cost and popularity of streaming video, having a public-facing policy is important. It helps to manage user expectations and is useful for library staff who are not directly involved in the acquisition of streaming but work directly with users such as public services staff who manage the library's chat service for troubleshooting purposes.

Licensing and managing streaming video content is complex and requires specialized workflows and practices that may not be appropriate to include in a public-facing CD policy but are important for library staff with selection and management responsibilities. Due to its complexity, streaming video requires a collaborative approach to collection development that takes into account the streaming video life cycle, which consists of acquisition, access, administration, support, and evaluation thus potentially including subject librarians, collection development, acquisitions, cataloging, and IT; therefore, it is best practice to develop selection criteria, considerations, or preferences as a group to ensure that topics such as licensing and copyright, platform features (e.g., accessibility), usage statistics, authentication and access, and discovery are covered. The criteria can be used to evaluate potential purchases, serve as a guide for discussions with content providers (e.g., provision of MARC, usage statistics), and be used as a training tool for subject librarians or any other library staff member who is involved with streaming video. Yale University Library has both a public-facing CD policy and internal guidelines for subject librarians and collection managers that outline criteria and procedures for the selection, acquisition, and access to streaming video content. The guidelines help subject librarians navigate the complexities of multiple content providers, an array of purchase models, copyright and licensing restrictions, and internal workflows and practices. Yale University Library's video collection development policy and internal streaming video selection guidelines are included in Appendix A.

Questions to consider when drafting a policy

- Why does the library license streaming video? (e.g., support teaching and research, preservation)
- How does streaming video fit within the library's broader collection development strategy and collecting philosophy? Does the library want to build video collections?
- What formats are collected? Why? (e.g., streaming only or hybrid collection; support for film studies program)
- Does the library license streaming versions for titles owned in a physical format? If so, when?
- What are the library's preferred acquisition models? Are certain acquisition models used for specific purposes? Are there acquisition models that the library does not use? To what extent will the library source individual films?
- How is video funded? Are there price caps? What is considered a reasonable price?
- Who is responsible for the selection and management of streaming video?
- What criteria does the library use to fill user requests?
- What criteria does the library use when making selection decisions? (e.g., relevancy, permitted use, platform features, picture quality, access and discovery)
- What streaming video information is relevant/useful to instructors and the general user community? (e.g., copyright restrictions, PPR, information regarding commercial streaming services)
- What is appropriate for a public-facing collection development policy? Should the library draft internal guidelines for library staff that provide more detailed selection criteria, and guidance on license negotiation, ordering, access, and discovery?

Selection Considerations and Evaluation

The goal of selection is to determine the types and kinds of resources and services the library should provide its users, and depending on the library's mission, the broader worldwide community (i.e., preservation of the cultural record) (Gregory 2019). Selection criteria are a core component of collection development policies that guide decision-making and explain why certain resources are chosen over others. Like other e-resources, factors to consider when licensing streaming video include content, cost and acquisition models, platform features and user experience, accessibility, availability of usage statistics, access, and discovery. Although these are important factors to consider when selecting content, the reality is that libraries often have limited choices available when licensing streaming. Because of these limited options, it is difficult to have concrete criteria or rules to evaluate resources. As a result, it may be more appropriate to develop selection considerations and preferences to guide the process.

Content

For most academic libraries, the resource's relevancy to the curriculum is likely the most important measure of value. If the resource is deemed to support the curriculum then it is acquired. However, unlike other resources that are selected for collection building purposes, much of the streaming video content libraries license is requested for courses or by faculty; therefore, there is less traditional selection of content, which raises long-term concerns regarding the curation of library streaming media collections. At a 2019 ALA Midwinter session hosted by the ALCTS Collection Management Section (CMS) Collection Development Librarians of Academic Libraries Interest Group and the Film and Media Round Table, Lisa Hooper, Head of Media Services at Tulane University, highlighted that in 2010 Tulane spent 5 percent of the media budget on faculty requests and the remaining funds to "strategically build a balanced collection"; however, in 2019, the library spent 95 percent of the budget on faculty requests resulting in noticeable gaps in the collection (Roll 2019, 409). Due to cost, librarians typically do not select streaming video titles on a title-by-title basis as they do with monographs. Instead, they must strategically choose acquisition models that align with the library's collecting priorities and preferences. Librarians are more likely to make selection decisions for subscriptions and perpetual streaming video collection purchases. It is important to note that librarians can decide which collections to subscribe to or purchase, but they often cannot curate or customize the content in those collections.

Selection Considerations

- Does the resource meet user needs or fill gaps in the collection?
- Does it build on collection strengths?
- Does it support an area of emerging research?
- Is it authoritative? Is it quality content? (e.g., filmmaker, producer)
- Is the content current? Important when evaluating content in streaming collections (subscribed or purchased). For example, outdated science or clinical content is problematic.
- Does the content duplicate content already in the library's collection? Before purchasing and subscribing to collections or at the point of renewal, it is best practice to do an overlap analysis.
- Do subscribed or purchased collections provide exclusive content?

Costs and Acquisition Models

Cost varies greatly by acquisition model and by content provider. The choice of acquisition model will depend on affordability, expected use (e.g., course reserves—short-term license), the library's preference for ownership or access, and available infrastructure (i.e., local hosting). Because of the variety of acquisition models, it is important for the library to establish preferred models, possibly by use case or expected use. Selection considerations for individual acquisition models touch upon the overall library collection development goals, content (e.g., core to a subject area), budget, access, and metadata management. These are a few general considerations when selecting an acquisition model. Selection considerations by acquisition model are addressed in greater detail in Chapter 2.

Selection Considerations

- What are the library's collection development goals—meet immediate needs, support distance education, collection building, etc.? What are the library's priorities?
- What is the content and how will it be used? (e.g., perpetual access model for content that is core to a discipline—performing arts)
- Can the library afford it? What is a reasonable price considering the library's collection development goals, priorities, and budget? What are the budgetary advantages and disadvantages? (e.g., subscription—expensive but predictable)
- Can the library manage it? (e.g., staffing, access, infrastructure, metadata)
- Can it be customized? (e.g., use-driven collections)

Copyright and Licensing

The license agreement outlines permitted use as well as other important terms and conditions, such as vendor notifications regarding platform migrations or when content is removed from a subscribed collection. Streaming video licenses should allow for educational, instructional, and classroom use. Large and educational streaming content providers usually have their own licenses, whereas small and foreign distributors and individual filmmakers may not. In these cases, a short streaming model license is useful, such as Yale Library's model streaming license located in Appendix B.

Selection Considerations

- How does the license define authorized users? How does the library define authorized users?
- Are there restrictions on access? Remote? International? Some licenses may specify that content must be accessed within the United States due

to international distribution rights, which is problematic for authorized users accessing the content while abroad (e.g., students studying abroad).
- Does the license address accessibility?
- Are public performance rights included?
- For perpetually purchased locally hosted content, does the library have the right to create copies of the file for backup or archival purposes? (This is a better option than life of file, which refers to perpetually licensed streaming content whereby the library's entitlement persists so long as the digital media file is viable).
- For perpetually purchased provider-hosted content, does the provider supply digital files for archival purposes?
- Does the provider respect user privacy? Does the provider gather personal user information? (e.g., creation of personal accounts).

Platform Features and User Experience

Although the library may provide users with a large and diverse amount of streaming video content, users will not have a positive experience if the platform is difficult to search and navigate; therefore, it is good practice to request a trial whenever possible.

Selection Considerations

- What search options are offered? Is it easy to browse content by subject, producer, etc.? Is it easy to find known films?
- Does the platform offer the ability to create clips?
- Does the platform offer the ability to create playlists/watchlists?
- Does the platform have shareable, stable URLs for individual videos?
- For short-term licenses, are license expirations displayed for each video?
- Does the platform offer a range of resolutions, applicable to varying broadband availability?
- Is the player reliable, and compatible with most browsers, and does it offer the ability to embed/link into websites/course management systems?
- Does the platform offer accessibility features?
- Can the library add branding?

Accessibility and Picture Quality

The Americans with Disabilities Act (ADA) requires public and private universities to provide equal access to postsecondary education for students with disabilities. For streaming video, the presence of captions, transcripts, audio descriptions, and player keyboard controls are important for accessibility purposes. Both captions and transcripts include spoken dialogue and all audio information, including sound effects and musical cues present in the video recording, but they are displayed differently (Spina 2021, 13). Player keyboard controls allow individuals with vision disabilities or mobility impairments to use the keyboard to play, pause, and sometimes forward and rewind videos rather than using a mouse.

Selection Considerations

- Are videos captioned and are audio descriptions available?
- Are video transcripts searchable, printable, or interactive?
- Does the platform have media player keyboard controls?
- Locally hosted: Does the content provider provide caption files?

Usage Statistics

Due to the high cost of streaming video, usage is essential for assessing whether the library is achieving a good return on investment (ROI). Usage statistics can help identify those resources with low usage that may require greater promotion or are not easily discoverable in the library's catalog or discovery layer. Usage data is also useful when making renewal decisions and in budget planning.

Selection Considerations

- Does the provider supply COUNTER-compliant usage statistics?
- What type of usage data does the provider offer?
- How easily can the library compare usage statistics across platforms to make renewal decisions?
- If the provider does not supply COUNTER-compliant data, do they provide a guide to compare their metrics with standard metrics?

Access and Discovery

Libraries cite cost as one of the biggest challenges to providing streaming content to their users, as well as their ability to keep up with demand and stay within budget; therefore, it is important that the content is used, findable, and accessible from multiple discovery points.

Selection Considerations

- Does the resource allow for simultaneous access by an unlimited number of users?
- Are there restrictions on remote access?
- For collections (subscribed or perpetually purchased), are there stable links to individual videos?
- Does the platform support IP authentication, single sign-on, etc.?

- Is the content accessible on a variety of devices?
- Does the platform or publisher require specific passwords to access content?
- Does the provider maintain an administrative portal to access usage statistics, MARC records, etc.?
- Does the provider supply MARC records?
- Locally hosted: What file format does the content provider supply and how is it delivered?

Conclusion

Demand and use of streaming video continue to increase for instructional, research, and recreational purposes. Users have become accustomed to instant access to a vast array of content, and streaming video has become an essential component of online instruction and distance learning. The evolution of the streaming video market, the high cost of streaming licenses, and the variety or lack of acquisition options are reshaping library collection development philosophies, transitioning from a focus on collection building to a service-oriented model; this model prioritizes providing access to meet the diverse information needs and preferences of users rather than building, curating, and preserving video collections. This shift is evident in the professional literature which rarely discusses "core video collections," but focuses on providing practical guidance on meeting current user needs in light of budgetary constraints, limited staffing capacity, and lack of the technological infrastructure to host streaming media. In fact, the content in this book is a reflection of that shift.

Streaming Vignette—Appalachian State University

Name of library: Appalachian State University
Reporting librarian and title: John Wiswell, Health Sciences Librarian, Associate Professor; Dr. Whitney Bevill, Humanities Librarian for Collection Management, Assistant Professor
Type of library: Public university, primarily undergraduate with some graduate and doctoral programs, located in Boone, North Carolina
User community: FTE of over 20,000 students
Total streaming budget: $62,500

Like subject librarians at many universities, Dr. Bevill liaises with faculty in the humanities when selecting collection materials in all formats, including streaming video. The collection management librarians at Appalachian work closely with the Resource Acquisition and Management (RAM) team, who provide support for acquisitions, cataloging, and electronic resources management. After collection management librarians identify resources to acquire, the RAM team handles sourcing and licensing of content. For larger packages, the coordinators of collection management and RAM may work on pricing negotiation, but individual titles are typically not negotiated beyond vendor pricing. Since streaming media licenses often offer differing term lengths, Dr. Bevill takes anticipated use into consideration. Perpetual licenses are frequently the preference, though short-term licenses of one or three years may be the only options, depending on the content provider.

Appalachian began a PDA program with Kanopy in 2021. As has been the case with many libraries, they found that allocated funds were spent more quickly than expected, and within a year, the plan was switched to mediated PDA. Additionally, there is also a plan with Swank where Appalachian prepurchases tokens for films and tops up throughout the year as needed. Other directly licensed collections include Alexander Street's American History in Video and Dance in Video, Docuseek2, and Naxos Video Library. Access is also provided to Films on Demand through NC Live, a statewide consortium of public and academic libraries in North Carolina. Appalachian does not typically license individual films for the collection from small or independent distributors, though the library has worked on getting PPR rights for individual films to be shown at a campus film festival.

Complex university policies for procurement also make negotiating license agreements with new content providers challenging and time-consuming, which is another reason why Appalachian relies on existing relationships with major vendors like Kanopy and Swank for their streaming needs. Cost-prohibitive titles are another challenge, particularly with more obscure or niche content, which can cost upwards of $500 or more for a single title. On occasion, Dr. Bevill has had to tell a faculty member that a title was prohibitively expensive; in most cases, faculty have been understanding and willing to consider alternatives. Similarly, larger packages like JoVE have proven to be challenging to negotiate at a price point that can be budgeted for sustainably.

For the most part, Appalachian's collection building for streaming video is done at the point of need as faculty members request titles for instruction. As such, little assessment of usage is done for the titles being acquired. Collection management is approached on a semester-to-semester basis, as title requests come in from instructors. Internally, a spreadsheet is used to track requests by semester, averaging around 150 titles per semester. Once requests are fulfilled, links to the titles are distributed to the requesting instructor. Records are also loaded into the catalog for streaming titles, even those which have short-term licenses.

Appalachian State University Library General Collection Development Policy, June 2023: https://library.appstate.edu/sites/library.appstate.edu/files/collection_development_general.pdf

Streaming Vignette—Delaware County Community College

Name of the library: Delaware County Community College Library
Reporting librarian and title: Andrea Rodgers, Associate Professor and Electronic Resources Librarian
Type of library: Community college, located in Delaware and Chester counties, Pennsylvania; six campuses
User community: approximately 7,600 for-credit students, more than 75 percent are part-time, and nearly 40 percent are enrolled in at least one distance education course
Streaming Video Budget: 5.5 percent of the e-resources budget

Delaware County Community College (DCCC) has six locations in the Philadelphia suburbs, serving residents in Delaware and Chester counties. Most library services are located on the Marple Campus within the college's Learning Commons, which also includes tutoring and writing services and a computer center. Andrea Rodgers, Associate Professor, and Electronic Resources Librarian, highlights that as a community college, teaching is at the forefront and informs how the library develops its collection and makes collection-related decisions. The library's collection development policy includes general selection and evaluation criteria and e-resource-specific guidelines, but it does not address streaming video as a separate format. DCCC users have access to the Films on Demand Master Academic Package, Docuseek2, Alexander Street's Theatre Performance and Design Collection, and Kanopy. Although the library manages the various streaming video collections, the Office of Online Learning pays for the Films on Demand and Docuseek2 subscriptions and negotiates those collections' pricing and licensing terms directly with the streaming providers. All other streaming video licensing and costs are negotiated by the Vice President for Finance and Administration's office. In these cases, the library makes renewal recommendations, but the final decision to renew is made by the VP of Finance and Administration, which Rodgers notes causes a certain level of anxiety but fortunately has not been a problem. The library's budget covers the subscription costs to Alexander Street's Theatre Performance and Design Collection and licensing to Kanopy films using a PPU model. DCCC does not do short-term or perpetual title-by-title licensing.

In 2018, DCCC started a Kanopy PDA program and worked with Kanopy to determine a reasonable budget, and although they had not actively promoted the collection, within a matter of months the budget was depleted. After identifying that most of the licenses triggered were filmed lectures in *The Great Courses* series—not documentaries and feature films to support the curriculum as initially intended—the library requested additional funding from the college administration and removed *The Great Courses* from the PDA pool which covered expenditures for the remaining portion of the academic year. In 2019, after a review of PDA usage and expenditure data, the library opted to move to a PPU model which would provide greater cost-savings. Rodgers provided a detailed account of the college's Kanopy experience in *College & Research Libraries News* (Rodgers 2019).

The library continues to use a PPU model for Kanopy videos and currently does not limit the number of views allowed per user. Rodgers highlighted that PPU works great for them and doubted that they would be able to offer Kanopy's impressive catalog without this model. It allows them to stay within budget, minimize the amount of staff mediation, and provides a good user experience (i.e., broad content and immediate access). Rodgers also noted that based on her experience, community college students differ from their residential campus counterparts in that they tend to search and use library streaming video resources more for curricular purposes rather than entertainment.

The library has no dedicated cataloging librarian; therefore, MARC records for streaming videos are not loaded into SirisiDynix, the library's integrated library system (ILS). However, Films on Demand, Docuseek2, and Theatre Performance and Collection Design are activated in Summon, the discovery layer. Unfortunately, Kanopy videos are not available in Summon and users must directly search Kanopy's platform.

Budgetary limitations and uncertainty impact the library's streaming collections strategy. As noted above, the library does not pay for or have final decision-making power over its largest streaming subscriptions—Films on Demand as well as Docuseek2—therefore, it cannot cancel/modify subscriptions and repurpose funds if needed (though they can make recommendations and the Office of Online Learning is receptive to them). For the past five years, the library also has had interim deans, and as a result, the budget request process and the budget itself have been in flux. Consequently, the library does not enter into multi-year agreements to contain costs and prefers to use deposit accounts with vendors to ensure that funds are available throughout the fiscal year to cover expenditures.

Assessment is important to support budget requests and maximize available funds. Rodgers runs annual COUNTER reports for all databases at the end of the academic year and keeps five years' worth of data. She looks at annual usage, creates graphs to show historical usage, and compares usage data by resource type (e.g., e-books and streaming video). To help inform renewal decisions, Rodgers uses usage and expenditure data to identify e-resources with high cost-per-use and those that are cost-efficient. The library has recently migrated to OCLC's Hosted EZproxy with a license to EZproxy Analytics, which transforms EZproxy data logs into a data management tool. It provides usage data across e-resource platforms and combines it with unique user identifiers at the point of authentication (e.g., course subject). Rodgers recently used EZproxy Analytics to see an uptick in drama students using the Theatre Performance and Design Collection, which was initially licensed to support English literature courses, but for which usage had been declining for some time. In general, she notes it is common to see peaks and valleys in usage depending on the courses being taught and the faculty teaching those courses. In the case of Theatre Performance and

Design, the faculty member who used the collection in their courses is no longer at the college.

For Rodgers, the main challenge for licensing streaming video is cost, but she is happy to see that streaming providers are rethinking their acquisition models to better serve the diverse needs of libraries of all types. The library would like to add new streaming content to support known teaching needs, but this will require a revaluation of current subscriptions and possibly cancellations to free up funds to cover the cost.

DCCC, Collection Development Policy: https://learningcommons.dccc.edu/collection-development-policy

Reference

Rodgers, Andrea. 2019. "Once upon a Time in Streaming Video: A Community College's Adventure with Kanopy's PDA Model." *College & Research Libraries News* 80 (9): 498. https://doi.org/10.5860/crln.80.8.498.

References

American Library Association. 2017. "The Library Mission." Accessed January 3, 2024. http://www.ala.org/tools/challengesupport/selectionpolicytoolkit/mission.

Association of College and Research Libraries. 2018. "Guidelines for Media Resources in Academic Libraries." Chicago: American Library Association. www.ala.org/acrl/standards/mediaresources.

Ballestro, John. 2019. "Business Models for Streaming Video." In Eric Hartnett (ed.), *Guide to Streaming Video Acquisitions*, 27–38. Chicago: American Library Association.

Brancolini, Kristine. 2002. "Video Collections in Academic Libraries." In *Video Collection Development in Multi-Type Libraries: A Handbook*, edited by Gary Handman, 41-69. Westport CT: Greenwood Press.

Carmichael, Michael, Abigail-Kate Reid, and Jeffrey D. Karpicke. 2018. "Assessing the Impact of Educational Video on Student Engagement, Critical Thinking, and Learning: The Current State of Play." White Paper. Sage. https://group.sagepub.com/white-paper-archive/assessing-the-impact-of-educational-video-on-student-engagement-critical-thinking-and-learning-the-current-state-of-play.

Carrico, Steven, Michelle Leonard, and Erin Gallagher. 2016. *Implementing and Assessing Use-Driven Acquisitions: A Practical Guide for Librarians*. Practical Guides for Librarians 23. New York: Rowman & Littlefield Publishers.

Cooper, Danielle, Dylan Ruediger, and Makala Skinner. 2022. "Streaming Media Licensing and Purchasing Practices at Academic Libraries: Survey Results." *Survey*. Ithaka S+R. https://sr.ithaka.org/publications/streaming-media-licensing-and-purchasing-practices-at-academic-libraries/.

Dixon, Jennifer. 2017. "The Academic Mainstream: Streaming Video is Becoming a Mainstay on College Campuses, but Discovery and Walled-off Content Create Their Own Challenges." *Library Journal* 142 (14): 42–5.

Ezor, Jonathan 2013. "Streaming While Teaching: The Legality of Using Personal Streaming Video Accounts for the Classroom." *Albany Law Journal of Science & Technology* 23 (1): 221–36.

farrelly, deg. 2014. "Streaming Video." In Becky Albitz, Diane Zabel, and Christine Avery (eds.), *Rethinking Collection Development and Management*, 215–29. Englewood: Libraries Unlimited.

farrelly, deg and Jane Hutchinson. 2014. "ATG Special Report: Academic Library Streaming Video: Key Findings from the National Survey." *Against the Grain* 26 (5). https://doi.org/10.7771/2380-176X.6852.

Francis Wilson Thompson Library, University of Michigan-Filnt. "Library Services and Resources for Faculty: Media in the Classroom and Copyright." Last modified: March 12, 2024. https://libguides.umflint.edu/faculty/media.

Fraser-Arnott, Melissa. 2022. "Exploring Public Library Identity Through Mission Statements." *Public Library Quarterly* 41 (3): 236–56.

Gregory, Vicki. 2019. *Collection Development and Management in the 21st Century*. Chicago: American Library Association.

Grombly, Amanda A. 2020. "Streaming Video Acquisition Models for Small to Medium Academic Libraries." *The Serials Librarian* 79 (1–2): 163–76. https://doi.org/10.1080/0361526X.2020.1772169.

Handman, Gary. 2010. "License to Look: Evolving Models for Library Video Acquisition and Access." *Library Trends* 58 (3): 324–34. https://doi.org/10.1353/lib.0.0094.

Hu, Brian. 2022. "Physical Media in the Age of Streaming: a View from the Film Studies Classroom." *Stream Magazine* Vol. 1. https://www.videotrust.org/stream-magazine/vol-01/physical-media-in-the-age-of-streaming.

Joint Digital Content Working Group, 2020. "The Need for Change: A Position Paper on E-Lending." American Library Association. https://www.ala.org/tools/sites/ala.org.tools/files/content/The-Need-for-Change-A-Position-Paper-on-E-Lending-by-the-Joint%20Digital-Content-Working-Group-2.pdf.

King, Diana. 2024. "Film and Video Collections: Commercial Home Video Streaming Platforms and Services." UCLA Library. Accessed March 12, 2024. https://guides.library.ucla.edu/videocollections/homevideo.

Levenson, Helen N. and Shawn V. Lombardo. 2023. "The State of Streaming Video Content at Academic Libraries." *Collection Management* 48 (4): 397–408. https://doi.org/10.1080/01462679.2023.2255561.

Lowe, Randall A., Fanuel Chirombo, John F. Coogan, Alexander Dodd, Corrie Hutchinson, and Judith Nagata. 2021. "Electronic Resources Management in the Time of COVID-19: Challenges and Opportunities Experienced by Six Academic Libraries." *Journal of Electronic Resources Librarianship* 33 (3): 215–23. https://doi.org/10.1080/1941126X.2021.1949162.

MacDougall, Ruby and Dylan Ruediger. 2023. "Teaching with Streaming Video: Understanding Instructional Practices, Challenges, and Support Needs." Ithaka S+R. https://doi.org/10.18665/sr.318216.

McCleskey, Sarah. 2022, "Film and Media Collections: Collection Development Policy." Hofstra University Library. Accessed January 5, 2024. https://libguides.hofstra.edu/Video/CollDevPol.

Netflix. n.d. "Educational Screenings of Documentaries." Help Center. Accessed February 3, 2024. https://help.netflix.com/en/node/57695.

Prelitz, Keri. 2022. "Revisiting the Streaming Video Patron-Driven Acquisition Program." *Journal of Electronic Resources Librarianship* 34 (4): 323–31. https://doi.org/10.1080/1941126X.2022.2133919.

Proctor, Julia. 2018. "Testing Assumptions: Is Streaming Video Really Preferable to DVDs?" *Journal of Electronic Resources Librarianship* 30 (2): 84–90. https://doi.org/10.1080/1941126X.2018.1465516.

Roll, Ann. 2019. "Report of the Session 'The State of Streaming Video in Academic Libraries,' Co-Sponsored by the ALCTS Collection Management Section (CMS) Collection Development Librarians of Academic Libraries Interest Group and the Film and Media Round Table. American Library Association Midwinter Meeting, Seattle, January 2019." *Technical Services Quarterly* 36 (4): 408–10.

Shaw, Wendy. 2011. "Collection Development Policies in the Digital Age." In Maggie Fieldhouse and Audrey Marshall (eds.), *Collection Development in the Digital Age*, 165–80. London: Facet Publishing.

Shirts, Peter. 2019. "The Library's Role in Providing Streaming Video." In Eric Harnett (ed.), *Guide to Streaming Video Acquisitions*, 92–7. Chicago: American Library Association.

Spina, Carli. 2021. *Video Accessibility*. Library Technology Reports, 57 (3). Chicago: American Library Association.

Stone, Graham and Jill Emery. 2013. "Investigation of New Content." *ALA TechSource* 51 (2). https://doi.org/10.5860/ltr.49n2.

Tanasse, Gisèle. 2021. "Implementing and Managing Streaming Media Services in Academic Libraries." *ACRL/Choice*. Retrieved from https://www.choice360.org/research/implementing-and-managing-streaming-media-services-in-academic-libraries/.

Valenti, Elizabeth, Tennille Feldbush, and Jean Mandernach, 2019. "Comparison of Faculty and Student Perceptions of Videos in the Online Classroom." *Journal of University Teaching and Learning Practice* 16 (3): 71–92. https://doi.org/10.53761/1.16.3.6.

Wahl, Mary. 2019. "Classroom Use." In Eric Hartnett (ed.), *Guide to Streaming Video Acquisitions*, 77–82. Chicago: American Library Association.

Willi Hooper, Michaela D. 2018. "Copyright for Movie Night: Film Screenings on Campus." *Journal of Copyright in Education & Librarianship,* 2 (1). https://doi.org/10.17161/jcel.v2i1.6576.

Wyatt, Neal. 2022. "Collection Rebalance." *Library Journal* 147 (7): 28–31.

2 Streaming Video Acquisition Models

Introduction

Since Film Media Group launched the first large streaming video collection for academic use in 2004, the number of library streaming providers and film distributors offering streaming media or the rights to stream their content for educational or institutional use has grown considerably, both in terms of quantity and variety of content (farrelly 2016). There are large streaming providers that license hundreds of titles accessible on their platforms such as Swank Motion Pictures, which specializes in films from major studios. Some providers are part of larger companies that license content in other formats such as e-books and databases like Films on Demand which is a part of Infobase or Alexander Street which is part of ProQuest. Some larger companies have integrated their content across platforms in various formats increasing the discoverability of video, such as ProQuest Academic One, a collection of four multidisciplinary products with a discovery layer that includes newspapers, databases, theses and dissertations, e-books, and Academic Video Online (AVON).

There are also independent distributors of all sizes that offer exclusive content available in physical format and streaming which often requires local hosting. A smaller number of films are available from independent filmmakers that license directly to libraries, and these often must be locally hosted. The number and variety of streaming providers and distributors that offer educational streaming rights and the various types of content they offer (e.g., feature films and documentaries) have resulted in complex and fluid acquisition models. To further complicate matters, the market is rapidly evolving due to mergers and acquisitions resulting in new business models, consolidation of content, and pricing models (Cooper, Ruediger, and Skinner 2022, 3). For libraries, this has meant an increasingly highly varied video acquisitions landscape at a time in which the demand for streaming video continues to increase for curricular, research, and recreational purposes, and most collections budgets are flat or not increasing at the rate of inflation (Bosch, Albee, and Romaine n.d).

The acquisition of streaming media is similar to the acquisition of e-books and e-journals. They are more expensive than their physical counterparts, can be licensed as part of packages or title-by-title, can be leased or owned, and may not be available for library acquisition (e.g., e-textbooks and test prep titles). However, unlike e-books and e-journals, the purchase of streaming media may also require a more boutique approach that includes sourcing titles directly from copyright holders, local hosting, and specialized workflows for ordering, invoicing, and records management.

Streaming media acquisition models can be divided into five broad categories that reflect the length of license (e.g., perpetual or short-term), hosting (e.g., provider platform or locally hosted), amount of content (e.g., single title or a collection), level and responsibility for selection and curation (i.e., library or user-initiated), and general availability. The "no streaming model" category captures films that are only available in physical formats, have no streaming option for institutional use, or the rights holder cannot be located or refuses to grant streaming rights (Ballestro 2019, 35).

Acquisition models:

- Perpetual models (single title or collection)
- Short-term models (single title or series)
- Subscription model (collection)
- Use-driven models (patron-driven acquisitions (PDA), evidence-based acquisitions (EBA), and cost-per-circ (or pay-per-use)
- No-streaming model

Although these categories are useful to conceptualize acquisition models, individual streaming provider and independent distributor models may not neatly fit into one of the above categories and often have provider-specific features and pricing models. Most large streaming providers offer multiple acquisition models on a single platform (e.g., Kanopy offers perpetual access licenses (PAL), individual title short-term licenses, subscribed collections, and patron-driven acquisitions), while others provide only a single model (e.g., Criterion Pictures USA—short-term licenses), and independent distributors often offer both perpetual and short-term licenses for the same films.

Like any other library resource, the selection of streaming media should align with the library's collection development policies and priorities such as the preference for e-resources for course reserves. Libraries should review and evaluate available acquisition models to ensure that they meet institutional pedagogical and research needs and that the library has adequate and sustained funding, staffing, and established workflows to license, purchase, manage, and provide access. The complex streaming acquisitions landscape, in conjunction with budgetary and staffing constraints, require libraries to develop a comprehensive video collection development strategy that addresses the role of physical and streaming media and includes a variety of acquisition models to meet immediate needs while ensuring long-term access.

This chapter provides an overview of streaming acquisition models, general use cases, and factors to consider when evaluating models including quality and variety of content, overall library collection development strategy, budget and cost, and implementation and management. Although there are references to licensing, budget planning, access, and discoverability, these topics are covered in-depth in other chapters of the book. A list of streaming providers and their acquisition models is available in Appendix D.

Perpetual Models

Although many films are not available streaming for perpetual purchase, the number of films available continues to grow. Distribution rights for films often change hands over time and streaming providers do not have perpetual distribution rights, which is especially true for popular feature films. Most films available for perpetual purchase tend to be documentaries, films purchased directly from independent distributors and filmmakers, and subject or thematic collections offered by larger providers such as Alexander Street's LGBT Studies in Video. A limited number of international and independent feature films are available for perpetual purchase from independent distributors such as Kino Lorber or directly from filmmakers. Many of these distributors license their titles directly to libraries as well as offer their content on third-party platforms. Unfortunately, major studio films are rarely available for perpetual purchase but a good number of films from Sony, Walt Disney Studios, and 20th Century Fox among others are included in subscribed collections (e.g., Swank Top 1000) or available as individual short-term licenses on provider platforms (e.g., Criterion Pictures).

Individual titles or collections that are purchased in perpetuity fall under two types of perpetual licenses: those hosted on provider platforms and digital site licenses (DSL) for the life of file. Providers such as Kanopy and Alexander Street offer perpetual access licenses (PAL) for videos and video collections hosted on their platforms for a one-time payment plus a hosting fee. They grant continuous access to purchased videos, but may not provide digital files for local hosting or archival purposes unless they file for bankruptcy or cease operation. In the case of DSLs, the library receives a file, usually an mp4, and uploads it to a platform that is managed locally. Some distributors often refer to "life of file" rather than "in perpetuity or perpetual streaming rights"; in such cases, the license grants streaming access for the life of file, and the license usually stipulates that the file cannot be replaced or migrated to a new file type without written consent or a renegotiation of the license. farrelly describes life of file "in the same way that a library owns a book, that is, as an artifact or container," the library does not own the content in the book but the book as an object (farrelly 2016, 6). However, it is best practice for libraries to try to negotiate the right to create copies of the file for backup purposes and convert the file format as deemed necessary; streaming video licensing is covered in detail in Chapter 4.

Single Title (Perpetual Access and Digital Site Licenses)

The cost for perpetual streaming rights, whether hosted on a provider platform or locally hosted, is expensive and varies widely by streaming provider. Streaming providers usually charge on average between $300 and $500 per title for a PAL and may charge a hosting fee per title or a platform fee. Hosting fees can quickly add up and should be negotiated before licensing the content. Some providers will waive hosting fees if the library maintains a preestablished spending level or maintains subscriptions to collections on their platform. Videos directly licensed from filmmakers and independent distributors which are locally hosted on average cost between $150 and $500 per title, plus the hidden labor costs required to receive the file or digitize a physical video, upload it to a platform, and catalog it. For those libraries that do not have the ability to locally host videos, some streaming providers offer hosting services for a fee, and some may offer hosting at no additional cost with a subscription to a streaming video package, such as Alexander Street with a subscription to AVON; however, some providers may limit the number of complementary titles hosted before the library incurs a cost such as Infobase that limits hosted titles to ten; additional examples can be found in Chapter 6. Although initially expensive, it may be more cost-effective to perpetually acquire titles that are heavily used rather than pay for a series of short-term licenses, which often occurs for titles regularly used in courses. In addition to the direct cost of purchasing a film, it is time-consuming to license individual films, especially if a title is locally hosted and sourced directly from a filmmaker or small distributor who has not previously worked with libraries. The hidden costs of licensing streaming media should also be taken into consideration when deciding to perpetually purchase a film.

The decision to purchase perpetual streaming rights is dependent on a variety of factors including expected use, available acquisition models, technical infrastructure, staff expertise (e.g., receive and upload files, digitize from physical formats), cost and funding, and the library's video collection development goals and priorities. A fundamental point to consider is why the library is pursuing perpetual streaming rights for individual titles which may include cost-savings over time for heavily viewed streaming titles, the transition from physical to streaming video for collection-building purposes, user request, or a perpetual streaming license is the only available purchase option. The most likely reason that most libraries purchase perpetual streaming rights is to meet curricular and research needs. Streaming offers users the ability to access content 24/7 regardless of location, which is not the case for physical media. Even before the COVID-19 pandemic, the use of streaming for curricular purposes was already common practice, and the pandemic solidified the pivot from physical to streaming reserves (Serrano and Fernandez 2023).

A perpetual streaming license may also be the only streaming option available, and possibly the only purchase option available if a physical video is not available for sale. Although expensive, purchasing individual titles in perpetuity will likely be an integral part of collection development for those libraries that aim to build and provide long-term access to media collections. At Yale, Area Studies Librarians started a pilot in 2022 to diversify the library's video collection by proactively acquiring DVDs of feature films and documentaries on approval from suppliers in non-western countries, as well as selecting and firm ordering films from those countries. However, they soon realized that most distributors and filmmakers only provided streaming files, and many only offered perpetual streaming rights. Librarians had not budgeted for the acquisition of life of file purchases, which were considerably more expensive than the amount they had budgeted for DVDs. They were also unfamiliar with available acquisition models and the cost associated with those models, as well as the internal order and licensing workflows for streaming videos. The pilot has led to a rethinking of current video collection practices, staff training, library workflows, and funding levels for video at Yale.

Collections (Perpetual Access)

Streaming collections available for one-time purchase are usually discipline-specific (e.g., performing arts), thematic (e.g., twentieth-century artists), within particular film categories or genres (e.g., television news), or profession-specific (e.g., clinical videos). Although they may be considerably more expensive than subscriptions, they provide long-term access to content that is often core to a field of study and an integral part of the overall general collection such as theatrical productions for the performing arts or silent films for film studies. Streaming video is often available for perpetual purchase as part of multi-format thematic digital collections such as ProQuest's Black Studies Center, which includes oral history interviews along with other primary source content. Although these are not exclusively video collections, video can play a prominent role in the collection and be one of the main reasons the library chooses to purchase the collection.

Purchased collections have both advantages and disadvantages. Unlike streaming video subscriptions, they provide stability of content and are easier to manage. Videos may be added to complete a collection but are rarely withdrawn unless the streaming provider loses distribution rights. Like purchased e-book collections or e-journal packages, the library should track entitlements to ensure that access is not lost. It is best practice for libraries to specify in the license agreement that the library is entitled to claim digital source files for backup or archival purposes. Although long-term stable access is an advantage, due to the number of videos in collections, it is probable that a subset may get few to no views. Furthermore, purchased collections on provider platforms often have annual hosting fees that may increase by 3 to 5 percent annually. The cost of the collection and hosting fees should be negotiated together, and when applicable, the library should use overall spend with the streaming provider as leverage to reduce or waive the fees. Another factor to consider when purchasing a collection is that some collections are released in parts or volumes, and unfortunately, libraries are usually not aware of forthcoming volumes at time of purchase and may find themselves continually trying to "complete" collections.

As noted earlier, although purchasing collections in perpetuity is expensive, an outright purchase is a good option if the content and format are core to a discipline (e.g., performance art, dance, theater), the collection is interdisciplinary and not licensable at the title level (e.g., short newsreels), or the library has one-time funds to spend and the collection is on the library's desiderata list.

Table 2.1 Factors to Consider: Perpetual Models

Content	• Does the content support the curriculum? • Is the content core to a subject area, discipline, or area of focus within the general collection?
Collection Strategy	• Why is a perpetual license preferable? • What are the library's collection development priorities—meeting immediate curricular needs, collection building, providing e-access for distance learners, etc.?
Budget and Cost	• Is it more cost-effective to acquire perpetual streaming rights for titles regularly used for teaching? • How will the library budget and plan for expensive perpetual access purchases? • Is it possible to pay in installments? • Vendor-hosted titles: How much are the hosting fees? Does the vendor waive fees and what are the requirements? Is there a cap on hosting fees?
Management	• Locally hosted and managed videos: Does the library have a content management platform to host video files? Does the library have the staff bandwidth to receive, upload, and manage the video files?

Short-Term Models

A 2022 Ithaka survey of academic libraries found that 94 percent of respondents purchase short-term streaming media licenses (Cooper, Ruediger, and Skinner 2022, 5). For the sake of simplicity, short-term licenses here refer to single titles or multi-part series licensed for a finite period, usually one to three years and occasionally five years. This content is usually requested by users, primarily used for teaching and sometimes institutional programming (e.g., student clubs), and hosted on provider platforms; however, a small number of educational distributors provide short-term licenses that must be locally hosted. Short-term licenses vary in cost, pricing models, order process, and overall management. They are managed at the title level rather than as a package and provide more library control over what is licensed and for how long. However, because license dates vary by title, it is difficult and time-consuming to monitor and update metadata for titles with short-term licenses.

Provider-Hosted Short-Term Licenses

Provider-hosted single limited licensed videos can cost on average between $100 and $300 per year with a lower cost per video if licensed as part of a series or multi-part collection. If the budget allows, multi-year licenses may be a better option for regularly used videos because the annual cost is lower, decrease the amount of staff time to license individual videos, and reduce the risk of users losing access to content when needed. A library can also negotiate better pricing when using a token-based system or a deposit account. For example, a library can propose to license a defined number of films in a year for a discount, and the discount increases as the number of licensed videos increases. In this case, the library can use data from previous years to estimate the number of videos to license.

Licensing and managing videos with short-term licenses can be time-consuming, but platform features and provider notifications can facilitate the process. The ability to license videos directly on the platform rather than contacting the content provider to submit an order substantially reduces staff time and reduces the time from user request to access. Managing license dates is also a challenge; once the license expires, the video is no longer accessible. Some providers send notifications when licenses are about to expire, while others require that the library monitor expiration dates in the admin module on the platform to ensure that videos used for teaching are accessible for the entire duration of a course. Some platforms have visible expiration dates on the public-facing interface which helps users, especially faculty who do not formally submit course reserve requests, to plan and ensure that their students have access when needed.

Locally Hosted Short-Term Licenses

Not all streaming content is available on provider streaming platforms and requires that the content be sourced directly from a distributor or filmmaker and locally hosted. Most distributors that focus on the academic market offer both perpetual and limited DSLs. Small distributors and filmmakers who have not worked with libraries may propose expensive limited DSL, and in these cases, it is helpful to reference established distributor licensing and pricing models. Cost is the main reason to opt for a limited license for locally hosted content when a perpetual license is available. The cost of locally hosted short-term licensed videos may be less expensive than life of file, but ensuring that videos are no longer accessible after the license expires is a challenge. Unlike videos hosted on provider platforms, locally hosted short-term access videos require the library to manually identify, remove, and delete titles when the license expires.

Table 2.2 Factors to Consider: Short-Term Licenses

Collection Strategy	• When does the library pursue short-term licenses? • When is a short-term license preferable to a perpetual purchase? • When should a video be licensed for one, three, or five years?
Management and Access	• Will records for short-term licensed titles be discoverable in the library catalog? • Vendor-hosted titles: Are expiration dates visible to users (important for faculty who bypass the course reserve request process resulting in student's losing access during the semester)? • Does the vendor notify the library before titles expire? • Locally hosted and managed titles: Does the library need to develop a workflow to manually identify, remove, and delete files at the end of the license period? If so, who will do this work? Does the library-managed streaming platform have features to facilitate this process?

Subscription Models

According to a 2021 *Choice* white paper on the implementation and management of streaming media services, most academic libraries use and prefer streaming subscriptions (Tanasse 2021, 15). An Ithaka survey the same year indicated that 90 percent of libraries use subscription models (Cooper, Ruediger, and Skinner 2022, 5). Streaming video subscriptions are renewable annual contracts that can act as aggregated databases in which the content is regularly updated, or as a collection with a fixed title list. They vary in size, cost, and stability and may coexist with other streaming acquisition models on the same platform. Subscriptions provide unlimited access to a large amount of content hosted on a provider platform at a fixed price. Unlike e-journal subscriptions, which often include post-cancellation access, streaming subscriptions do not. Once the subscription period is over, the library will need to renew the subscription or access to the content will be lost. Some streaming subscriptions include thousands of videos, and it is important to evaluate the breadth and depth of coverage to determine if the content is of value to the library's user community. It is likely that a significant number of videos will get little to no use, but the question is whether there is enough "quality" content to justify the subscription. Furthermore, the amount of exclusive content in packages varies, and before subscribing, it is useful to do an overlap analysis to determine duplication with content already licensed by the library.

Large subscription packages act like aggregated databases consisting of thousands of videos, and because of the number of distributors represented and the size of these collections, content providers regularly lose and gain distribution rights, as a result, content is added or dropped throughout the subscription period. The amount and type of content dropped should be considered before renewing and when negotiating renewal pricing. For example, if one of the reasons that the library decided to subscribe to a package is the availability of PBS titles, and many of these titles were dropped during the subscription period, the library should consider if it is worth renewing and at what price. Ballestro recommends that libraries treat streaming subscriptions like e-journal packages and monitor removals and transfers (Ballestro 2019, 30). Due to the size of some of these packages, it can be difficult to regularly monitor titles, but the library can take a strategic approach by monitoring specific distributors, filmmakers, or subject areas of interest, as well as overall number of videos in the package. Although most streaming providers send removal notifications, there is always the chance that titles currently used in courses may unexpectedly be dropped; therefore, libraries should have workflows in place to identify films that are dropped from subscriptions within the course reserve system which requires staff time and does not guarantee that the faculty member will be notified before students are expected to view the film.

In addition to large streaming subscriptions, libraries can also subscribe to curated collections that may focus on a specific area, producer, or bring titles together that are deemed of interest to academic or public libraries. This content may be available in larger subscription packages or may be exclusive to the collection. If large streaming video subscriptions are financially prohibited, smaller more focused collections may be a viable option. Some streaming providers offer curated collections from various producers or distributors that focus on film categories or genres (e.g., video art, documentaries, or feature films), subject areas (e.g., dance or music), interdisciplinary areas (e.g., LGBT studies), themes (e.g., global issues), or professions (e.g., counseling). There are distributor- and producer-specific subscriptions such as Kanopy's Criterion Collection or Alexander Street's PBS Video Collection. Other streaming providers offer subscriptions to exclusive discipline- and profession-specific content of interest to targeted audiences, especially in the social sciences, sciences, and medicine, such as Psychotherapy.net, Sage Video, and Journal of Visualized Experiments (JoVE). Streaming providers may also offer subscriptions to curated collections that the provider defines as essential or the most requested films on the platform such as Swank's Top 1000 and Kanopy BASE for academic libraries or Kanopy PLUS for public libraries.

Table 2.3 Factors to Consider: Subscriptions

Content	• Does the content support known instructions and research needs? • Does the collection include content appropriate for your user community? • Does the collection include a variety of quality content? • Does the collection include titles for which currency is important (e.g., medical and health titles)? If so, is the content current? • Does the subscription offer exclusive content? • Does the content duplicate physical holdings? Does the content duplicate library-licensed content on other streaming platforms or locally hosted?
Collection Strategy	• Considering expected use, content, and cost, does the provider offer other acquisition models that better suit library needs (e.g., license subcollections, license individual videos as needed)?
Budget and Cost	• Does the quality and quantity of content justify the cost? • Does the provider offer better pricing for multi-year commitments? • What is the annual increase?
Management and Access	• Does the provider provide MARC records? • Does the provider send notifications for new or withdrawn titles? • Are withdrawn titles available for perpetual purchase, and if so, is this information included in removal notifications?

Use-Driven Models

Use-driven acquisitions (UDA) is an umbrella term that incorporates multiple models and applies to any method of collection development in which content is acquired based on patron discovery and use rather than the anticipation of patron use (Carrico et al. 2016, xiii).

Patron-Driven Acquisitions or Demand-Driven Acquisitions

In a PDA model, also referred to as demand-driven acquisitions (DDA), users play a central role in the collection development process. Users are provided a large and broad array of content and the library licenses content that is used without the need for library staff mediation. Instead of the library selecting content in anticipation of user need, often described as a "just-in-case" model, PDA enables acquisition at point of need or "just-in-time." PDA emerged in the early 2000s with e-book aggregators such as NetLibrary, and was seen as a way to provide a greater choice of content while providing a good return on investment

(ROI) (Polanka and Delquié 2011). Streaming video PDA programs focus on access and are not meant to build video collections in a traditional sense (e.g., ownership). However, usage data is helpful in making acquisition decisions when a perpetual license is available at an acceptable price.

Streaming video PDA programs are continually evolving, and it is important to stay abreast of provider-specific changes to better utilize available funding and improve user experience and access to content. In fact, by the time of this publication, streaming video providers will certainly have tweaked their acquisition models to reflect changes in user behavior and library needs. Streaming video PDA programs vary in the number and types of videos available, license triggers (e.g., number and length of views that result in a purchase), level of curation, cost per video, payment options, and other features such as previews which may reduce the number of triggered licenses by users who are browsing the platform. Before deciding whether to pursue a PDA program, it is important to consider the role and purpose of the program within the library's video collection and available funds to cover the cost. Streaming films are extremely popular and are heavily used for entertainment purposes, therefore developing a strategy is key to ensure that the library stays within its budget.

Customization of the PDA collections can vary from no curation (e.g., entire catalog) to broad categories such as subject, distributor, audience, language, or title-level curation that excludes licensed content already available in the library's collection. Depending on the streaming video provider, the library can self-curate a collection based on local usage data and curricular needs or work with the streaming provider to curate the collection (e.g., Kanopy Smart PDA and Swank Flex Collection). The ability to highly curate a collection may initially be time-consuming but in the long run can help contain costs. Licensing individual titles as part of a PDA program is expensive, and it is important to limit a user's ability to trigger licenses for titles already licensed by the library whether leased or owned. Although records for streaming videos may be discoverable in the library's catalog, several studies have shown that both instructors and students tend to use internet sources such as Google as their primary means to discover streaming video content and not the library's catalog (Lohmann and Frederiksen 2018; Beisler, Bucy, and Medaille 2019; Wang and Loftis 2020).

Depending on the streaming provider, a deposit account may be required or recommended to start the program. Streaming providers can work with the library to determine a realistic budget based on factors such as the type and size of the institution, the length of the program, and the level of curation. A deposit account allows the library to better track spend, and if necessary, suspend the program when spending targets have been reached. However, some streaming providers will also invoice as videos are triggered or bundle videos into monthly invoices which may be a better option for libraries that cannot have deposit accounts due to administrative policies.

Once the provider has activated the collection, users can view videos and trigger licenses based on the number of views and/or length of each view. For example, Kanopy video licenses are triggered when a video is viewed four times and each view is more than thirty seconds long. The length of an individual license varies by provider, and some offer multiple options from one to three years which is determined before implementation. Once a license expires, the video goes back into the PDA pool and the process starts all over again. Alexander Street's DDA program is unique in that the library gets automatic perpetual access to qualifying titles when the cost for triggered licenses for a particular title has reached $499.

PDA programs offer a series of advantages including immediate access to content that libraries would not otherwise be able to afford, guaranteed usage, and limit the amount of staff time spent on selection and acquisition. Some streaming video providers offer PDA and PAL, and data from the PDA program can inform purchase decisions. However, there have been numerous publications in recent years that illustrate the challenges in developing, managing, and funding streaming video PDA programs, including the ability to adequately and predictably budget for the program, manage user expectations, and accurately maintain current MARC records (Lowe et al. 2020; Urban 2022; Foley 2019). Because of larger than anticipated use and spending, some libraries have turned off their PDA programs and moved to a mediated PDA model, in which library staff review and approve user video requests based on preestablished selection criteria. Requests are usually submitted via the streaming platform and designated staff receive email notifications when a title is requested. They can then license the title or reject the request and notify the user. In this approach, the user initiates the acquisition, but the library makes the final decision. Mediation enables the library to provide content that will be used while controlling spending. It also offers the library the opportunity to choose multi-year licenses for titles with expected high use (e.g., regularly used for teaching), which is less expensive over time. However, depending on the number of requests, mediation may require a significant amount of staff time and does not provide immediate access to the content.

Table 2.4 Factors to Consider: Patron-driven Acquisitions and Demand-driven Acquisitions

Content	• Does the provider offer content appropriate for your user community? • Does the provider offer titles in areas for which currency is important (e.g., medical and health titles)? If so, are the titles current?
Collection Strategy	• What are the advantages and disadvantages of expanding access to a provider's entire catalog or limiting access to certain producers, genres/categories of film, or subject areas?
Budget and Cost	• How will the program be funded? • Is a deposit needed? What happens to the deposit if it goes unspent? What are the invoicing options? Can the library track the balance in the provider admin portal? • What's the cost per video? How is cost determined—for example, the length of a video, part of a series, or by the producer? • Do individual videos in a series each trigger a purchase?

Table 2.4 (Continued)

Implementation	• Is it preferable to do a pilot before making a decision (e.g., 3 months)? • How long is the program (i.e., length of commitment)? • How and to what extent can the library customize the collection? Will the provider facilitate customization? What data or information are needed to customize the PDA collection? • For triggered titles, what is the default license period (e.g., 1 or 3 years)?
Management	• If customization is an option, who will curate the collection, monitor usage and expenditures, and provide overall oversight of the collection? • How will the PDA program impact standard ordering, invoicing, and discovery workflows? How will the library manage duplication of records across streaming collections—PDA, subscription, and owned? • How will the library decide when it is time to "turn off" the PDA program?

Evidence-Based Acquisitions (EBA)

In an EBA model, the library pays an upfront fee, and is given access to a pre-selected package or collection of streaming titles for a limited time period (e.g., four to twelve months). MARC records are loaded into the library catalog or activated in the discovery system. At the end of the pre-defined time period, usage data is analyzed to select a subset of titles against the fee paid for perpetual access. Streaming provider EBA models vary, and the upfront fee may be referred to as a participation fee, a deposit, or a subscription with a set number of perpetual access titles, but they are fundamentally the same. An EBA model focuses on collection building while providing access to a large amount of content that the library would not otherwise be able to afford. An EBA program can be customized to align with the library's long-term video collection development goals including replacing heavily used physical media with streaming, perpetual access to titles routinely used for teaching, diversifying the collection, or adding titles in expanding areas of scholarship.

To begin an EBA program, the library and streaming provider negotiate a fee which is paid upfront and can be significant based on the number of accessible titles and the duration of the program. In the case of an academic library, the fee may be based on the number of accessible titles, full-time student enrollment, or any other criterion a provider uses to price subscription-based e-resources. The cost per video may vary and should be discussed with the streaming provider before starting the program, as well as the usage metrics provided (Spratt et al. 2017). The next step is to set up the EBA collection which can include the content provider's entire catalog or a customized collection of titles. Depending on the streaming provider, this can be done at the title level, by distributor/producer, genre, subject, or collection (e.g., Alexander Street's Asian American Video). When the EBA program is activated, users have access to all videos in the library's curated collection. During this period, the library should track activity, review usage reports, and begin to discuss how the metrics provided will inform purchase decisions. At the end of the program, the library selects titles to purchase equal to the participation fee.

EBA programs strike a balance between user-initiated selection and library selection based on data and librarian expertise. They provide access to a significant amount of content and the library is able to select and own a portion of that content. Some EBA programs require upfront costs that may be too large for some libraries; in these cases, libraries should discuss the availability of smaller EBA programs with providers such as subject or producer collections which may be more affordable.

Table 2.5 Factors to Consider: Evidence-Based Acquisitions

Content	• Does the provider offer content appropriate for your user community? • How does the EBA collection complement the library's video collection? • Does the collection include content that the library would like to own?
Collection Strategy	• Does the program provide increased access to streaming video while also helping to build a collection that aligns with the library's video collection development goals?
Budget and Cost	• How will the program be funded? Can funds be reallocated to cover the cost of the program (e.g., reallocate funds from physical media acquisitions)? • Can the library track expenditures and/or titles selected in the vendor admin portal?
Implementation	• How long will the program run? • How will the library manage records? Will the library de-duplicate records for content already licensed?
Management and Access	• Who will be responsible for selecting titles for perpetual ownership? • What usage data does the vendor supply and is it sufficient to make informed purchase decisions? • What will be the workflow for selection and entitlement control? • How much time does the library have to make purchase decisions? Is there a date by which the library must make purchase decisions? What happens if the library does not select films for purchase by the deadline?

Pay-Per-Use

Pay-per-use (PPU) is a model usually found in public libraries in which the library pays a fee, often ranging from $2 to $6, each time a video is "checked out." To contain costs, libraries have developed streaming video-specific circulation policies that limit the number of videos a user can view/checkout during a preestablished period (e.g., five videos per month) as well as the number of holds allowed per user. Once a user has selected a video to view, they have a limited amount of time in which to view the video (e.g., three days), and if they choose to view the video after the allotted time period, they will use another allowed view or checkout. Depending on the streaming provider, users often must download applications to enable and access the content via their computers, smartphones, tablets, or television sets (Adamich, 2021). Some content providers, such as Hoopla, offer media packages that include streaming video, e-books, audiobooks, comics, magazines, and streaming music, all of which have similar circulation policies and offer a different user experience than video-only packages.

No-Streaming Model

Although the number of educational streaming video providers and streaming films available to libraries has increased in the past years, libraries will likely get requests for films that are not licensable for institutional use, lack streaming availability, or the rights holder cannot be located or refuses to grant streaming rights. There are films only available streaming on direct-to-consumer platforms as part of personal subscriptions or individual film rentals. Some of this content may be available in physical formats that can be added to the library's permanent collection; however, an ever-growing number of films and series are produced and distributed by streaming services such as Netflix or Hulu and are not available for library purchase in any format—streaming or physical. Although not ideal, users can pay to access content on direct-to-consumer platforms if needed for courses or research. In addition to personal use, Netflix allows one-time educational screenings for Netflix account holders for documentaries that display a Grant of Permission or an Educational Screenings Permission (ESP) on their details page. Netflix's description of one-time is fairly vague in that "you can't hold screenings several times in one day or one week but if, for example, you're an educator who wants to show these films or series once a semester over multiple semesters, that's okay" (Netflix n.d.). Some requested films may also not be available on direct-to-consumer platforms in a library's region. Distribution rights vary by country and region (e.g., North America), and although a German television miniseries may be available on Amazon Prime Germany, it may not be available on Amazon Prime in the United States or any other US streaming platform.

Table 2.6 Summary of Acquisition Models

Acquisition Model	Definition	Advantages	Disadvantages
Perpetual—perpetual access license	Perpetual access to a video hosted on a vendor's platform	• Ownership • No local server or hosting service	• Expensive • Hosting fees
Perpetual—digital site license	Purchase of video for Life of File and locally hosted	• Ownership • No hosting fees	• Expensive • Local server or hosting service • Staff time
Short-term license—single title, multi-part series (vendor-hosted)	Single title or series licensed for a limited time accessible on a vendor's platform	• Library control • Less expensive than perpetual purchase • No need for a local server • Short time from the point of license to access	• Leased • Expensive long-term for regularly used titles • Staff time
Short-term license—single title, multi-part series (locally hosted)	Single title or series licensed directly from the rights holder and hosted by library for a limited time period	• Library control • Less expensive than a perpetual purchase	• Leased • Local server needed or hosting service • Expensive long-term for regularly used titles • Staff time • Specialized workflow to ensure titles are removed at end of license period

(Continued)

Table 2.6 (Continued)

Acquisition Model	Definition	Advantages	Disadvantages
Subscription—package and collection	Subscription to a package or defined list of titles	• Predictable cost • High volume of content • Easier records management	• Expensive • Unstable content (large packages) • High percentage of content with little to no use
Patron-driven acquisitions	User triggers a limited license with no library mediation	• User initiates purchase and content is used • Immediate access	• Leased • Triggered content may not align with library collection development priorities • Difficult to control costs
Evidence-based acquisitions	Library makes perpetual purchase decisions based on usage data	• Ownership • Selection based on data • Defined title list	• Large upfront costs • Staff time (setup and data analysis) • Possible inconsistent usage data
Mediated	User initiates a purchase request and library approves request resulting in a limited license (e.g., 1 year)	• User-initiated requests • Library control over content and budget	• Staff time • Delay in access
Pay-per-use/Cost-per-circ	Fee paid each time a video is viewed (viewer can see multiple times during the preestablished checkout period)	• Opportunity to offer a wide range of content and control circulation	• Expensive • Users have a limited number of checkouts or views (most cases)

Conclusion

The academic streaming video market is fluid and evolving and reflects the general academic publishing landscape in which mergers, acquisitions, and the consolidation of content are commonplace leading to new business and pricing models. For example, ProQuest acquired Alexander Street in 2016 along with its extensive video catalog including AVON. In 2019, ProQuest rolled out ProQuest Academic One, a large subscription package that includes ProQuest Central (40+ databases), Academic Complete (200,000+ e-books), ProQuest Dissertation & Theses Global, and AVON which are all cross-searchable. Finally, Clarivate acquired ProQuest in 2021 which may affect the company's future video offerings, integration with Clarivate products (e.g., Web of Science), and pricing models. OverDrive, a content provider that licenses primarily trade e-books and audiobooks, acquired Kanopy the same year, which led some to ask if the purchase was an attempt to become "a more complete media solution for libraries," especially for academic libraries (Cooper and Schonfeld 2021). In the two years since the merger, OverDrive and Kanopy's platforms remain separate, but the company has an integrated marketing strategy linking its two platforms. OverDrive's expansion into the academic sector is perhaps most evident with the addition of academic publisher content including open-access e-books from Taylor and Francis and Wiley. It remains to be seen if OverDrive will integrate its platforms and offer multi-format packages (i.e., video, e-books, or audiobooks).

Streaming Vignette—Cranston Public Library

Name of the library: Cranston Public Library (CPL) and Ocean State Libraries Consortium (OSL)
Reporting librarian and title: Edward Garcia, Library Director, Cranston Public Library; Lisa Sallee, Assistant Director, Ocean State Libraries Consortium
Type of library: Public library system, located in Cranston, Rhode Island; Main library and five branch libraries
User community: Approximately 82,500 residents
Total streaming budget: $12,000 ($7,000 Kanopy deposit and $5,000 Hoopla deposit)

Ocean State Libraries Consortium (OSL) is a consortium of public libraries serving the Rhode Island community through statewide licensing of electronic resources, a shared catalog, and borrowing privileges at member libraries. Although OSL does not currently license streaming video content, it does facilitate the discovery and shared access to DVDs and Blu-rays via the consortium's shared catalog. A subset of the forty-eight member libraries has direct subscriptions to Kanopy and/or Hoopla, and according to Lisa Sallee, Assistant Director at OSL, it is not surprising that libraries with subscriptions are usually earlier adopters of new technology and are often located in urban areas. Because of the cost and backend management of streaming services like Kanopy, OSL is not considering consortial agreements for streaming video content at this time.

Cranston Public Library (CPL) comprises six locations and is situated in Cranston, Rhode Island, the second-largest city in the state. The library has taken a holistic approach to providing access to video by acquiring DVDs and licensing streaming video content from Kanopy and Hoopla. All non-consortial e-resource licensing and pricing is negotiated by Ed Garcia, the Director of the Library, who also manages the library's Kanopy and Hoopla plans. Community members requested that the library subscribe to Kanopy and Hoopla, and in the case of Hoopla, community members became aware of the service because it was available at other Ocean State Libraries. Garcia noted that each streaming service fills a different function: Kanopy provides more feature films and documentaries, while Hoopla offers more quantity and content types in the form of audiobooks, e-books, and streaming media. Kanopy offers good content curation and public performance rights (PPR) for some of its content which is of particular interest to the public schools in the area. Although Hoopla offers a large amount of content, Garcia highlighted content moderation as an area of concern, and some of the content is questionable and would not normally be purchased by the library. Furthermore, he has recently heard that Hoopla is adding more audiobooks with machine-generated narration which community members do not like.

In 2018, CPL began to license Kanopy videos with funds from the Friends of the CPL using a PPU model, in which each patron is allowed six play credits per month, and the library pays $2 to $5 per credit depending on the film, except for Kanopy Kids titles that do not require credits and have unlimited plays. Once a user has selected a video and used a play credit, the user usually has seventy-two hours to view the video. Garcia was recently notified that Kanopy is moving to a new ticket model, in which the library pays $1 per ticket, users are given a fixed number of tickets per month, and individual videos can range from two to five tickets. In essence, the new system allows libraries to set a cap on what users can "spend" rather than the number of videos watched. To ensure sustainable funding, Kanopy is now paid out of the library's operational budget and approximately $7,000 is spent per year; this amount is added to a deposit account at the beginning of the fiscal year and Kanopy supplies monthly deposit status notifications. The initial Kanopy budget was $5,000 in 2018, but in 2020, an additional $2,000 was added due to increased content usage. Because The Great Courses are generally more expensive than other videos in Kanopy's catalog, they were removed from the collection to help control costs. Although Kanopy continues to be a popular resource, use has plateaued, and Garcia does not expect to allocate additional funds in the foreseeable future.

Hoopla was added in 2020. The library sets a price cap for all material types and anything above the cap is not discoverable on the platform. In general, Hoopla usage has always been lower than Kanopy, but this has recently started to change now that Hoopla content is discoverable in Aspen, the new OSL discovery layer, which can be customized at the library level. OSL policy requires that all digital content in the shared integrated library system (ILS) be accessible to all state public library card holders; therefore, historically Kanopy and Hoopla content was not discoverable in the shared catalog. Although Hoopla's collection includes movies and television series, usage data clearly shows that audiobooks are used far more than any other content type and are driving the overall increase in usage.

Garcia does not expect the library to add new streaming video resources. Print circulation has not gone down, and therefore, it is difficult to repurpose funds for e-resource acquisitions. Although the library is acquiring fewer DVDs simply because fewer films are available on DVD, video game acquisitions have increased due to high demand. The city is also not in the best financial shape, and although the library's budget has not been cut, Garcia prefers to be cautious and ensure that the library's operating budget has sufficient funds to cover both collections and unexpected facilities expenses such as increased utility costs.

Regarding ongoing challenges to acquiring and managing streaming, Garcia identifies cost, content availability, and content moderation as being the most pressing. Current and evolving purchase models available to public libraries can be complex and expensive depending on the population served. According to Garcia, it is also worrisome that a significant amount of video content is not available for library purchase and is exclusive to commercial streaming providers. Furthermore, as libraries license more streaming video collections/packages, more needs to be done to ensure that libraries are getting access to quality content and not simply a lot of content.

Cranston Public Library Collection Development Policy: https://www.cranstonlibrary.org/policies/collection-development-policy/

Streaming Vignette—Dickinson College

Name of the library: Dickinson College (Waidner-Spahr Library)
Reporting librarian and title: Theresa Arndt, Associate Dean for Collections & Resources
Type of library: Private Liberal Arts College, located in Carlisle, Pennsylvania
User community: Full-time equivalent (FTE) of 2,100 students, primarily residential undergrads
Total streaming budget: Approximately 3.5 percent of total materials/services expenses

Dickinson College has a number of streaming video collections from major content providers such as Swank, a Kanopy Smart PDA plan, a Films on Demand subscription from Infobase, the Docuseek2 collection, and a subscription to Alexander Street's Academic Video on Demand that was licensed through PALCI, a consortium of academic and research libraries located in Pennsylvania and surrounding areas. As a smaller institution with a limited budget and staffing, the majority of streaming video content is acquired for curricular support. When included in the license, PPR notes are added in library records so that clubs or organizations may also make use of them. Licenses for large campus screenings are arranged for and paid for by other college offices, not the library.

In addition to subscriptions, Dickinson pursues streaming video content through other acquisition models. For Swank, prepaid tokens for individual films are purchased and titles are chosen on an as-needed basis to fulfill faculty requests. The Kanopy plan was initially unmediated PDA, which quickly ate up budget allocations. Arndt notes that at the time, much of the external publicity surrounding Kanopy gave the impression that all of the content was "free" to users. At the same time, it was recognized that the platform was very popular at Dickinson, so they were hesitant to drop it altogether. Kanopy's introduction of the Smart PDA option offered the opportunity to continue providing the content that users were accustomed to through a scaled-back model that allowed Dickinson to control costs. Working closely with their Kanopy sales representative, Dickinson curated an initial pool of titles based largely on their own historical usage of content on the platform. In addition to short-term licenses through Smart PDA, Dickinson will occasionally perpetually acquire Kanopy titles that have been licensed multiple times. The Kanopy BASE collection was also evaluated, however, it was determined that only 16 percent of Kanopy requests could have been fulfilled by the collection, per reports that the sales representative generated. Kanopy has also begun flagging content that supports DEI goals in collection building. Arndt cites the customer service and assessment metrics provided by their rep as being particularly helpful in arriving at a cost-effective acquisition model.

Film requests are received by the library through an internal form—managed in SpringShare's LibApps suite—that is filled out and submitted by faculty members. Requests can come across various subject disciplines; although Dickinson does offer film studies courses, the use of film tends to be cross-curricular. Arndt oversees a fairly small staff of four; the request form is typically first received by an acquisitions technician who will see if a title is already included in their existing collections, before checking to see if it can be sourced from a major supplier. If not, they will then investigate whether there is any supplier who can provide an institutional streaming license. In some cases, titles licensed from smaller distributors will be licensed and hosted locally on Dickinson's instance of the Panopto platform. If a requested title is available through institutional licensing and is not prohibitively expensive, it will be acquired in the streaming format. If not, a DVD copy will be sourced. While physical format acquisitions are declining overall at Dickinson, physical reserves for DVD materials are still used on occasion. Because the campus is residential, the assumption is that students will be able to pick up physical reserve materials if necessary when DVD is the only viable acquisition option. Some management of user expectations is required, as the majority of students—and faculty as well—prefer streaming video.

Dickinson utilizes Alma for their LSP with a Primo discovery layer. For subscription-based streaming collections, records are activated in the Alma Community Zone and made discoverable to users. Permanently acquired titles that are locally hosted or on DVD will get cataloged and made discoverable as well. For locally hosted titles, copy cataloging is typically done, sourced from an OCLC or LoC record. On rare occasions original cataloging will be done, such as cases where a source record does not exist for a title like a more obscure documentary or some non-English films. Additional local notes are added to the record, to denote any license restrictions or if PPR is available for a title. PPR availability for streaming collections is also indicated on the library's A-to-Z List. Short-term licensed titles do not get discovery records; instead, a link for the content will be sent directly to the requesting faculty member. Additionally, titles in the Kanopy PDA pool are not made discoverable through Primo either, though users can surface them directly on the Kanopy platform. Again, the limited staffing at Dickinson largely influences the policies on cataloging and discovery, as continually managing titles with license terms expiring throughout the year would be a heavy drain on staff bandwidth.

For assessment, Dickinson does look at COUNTER data for their larger subscription packages, which typically see good usage across the board. Kanopy has been offering bespoke reporting on a quarterly basis which does a good job of demonstrating the value of the Smart PDA plan. A particularly helpful assessment tool has been the request form; through the backend of the SpringShare admin site, a spreadsheet can be generated of all the received requests. This running history of requests can be used to address a number of assessment questions, such as, what is the request fulfillment rate? Who are the most frequently used content providers? How many repeat requests are made for a specific title? Knowing that a certain title is being requested repeatedly can inform collection decisions on whether a title should be acquired permanently.

Turning to ongoing challenges posed by streaming video collections, Arndt cites the overall lack of permanence for the format, particularly the fluidity with which individual titles drop on and off of platforms. Complex factors such as distribution rights can contribute to film content being made unavailable. As an example, she notes *Eyes on the Prize*, a landmark documentary series on the history of the civil rights movement in America, which was tied up in copyright issues regarding archival footage and music clearance, making it unavailable in any format for a prolonged period of time. Despite the continued growth of streaming video, not every title will be available, requiring libraries to sometimes rely on previously acquired physical format copies.

Waidner-Spahr Library, Collections Guidelines: Films: https://libguides.dickinson.edu/collections/films

References

Adamich, Tom. 2021. "ON-DEMAND STREAMING MEDIA-Considerations for Public Libraries." *Computers in Libraries* 41 (10): 18–21.

Ballestro, John. 2019. "Business Models for Streaming Video." In *Guide to Streaming Video Acquisitions*, 27–38. ALCTS Monograph. Chicago: American Library Association.

Beisler, Amalia, Rosalind Bucy, and Ann Medaille. 2019. "Streaming Video Database Features: What Do Faculty and Students Really Want?" *Journal of Electronic Resources Librarianship* 31 (1): 14–30. http://doi.org/10.1080/1941126X.2018.1562602.

Bosch, Stephen, Barbara Albee, and Sion Romaine. n.d. "Costs Outstrip Library Budgets | Periodicals Price Survey 2020." *Library Journal*. https://www.libraryjournal.com/story/Costs-Outstrip-Library-Budgets-Periodicals-Price-Survey-2020.

Carrico, Steven, Michelle Leonard, and Erin Gallagher. 2016. *Implementing and Assessing Use-Driven Acquisitions: A Practical Guide for Librarians*. Practical Guides for Librarians 23. New York: Rowman & Littlefield Publishers.

Cooper, Danielle, Dylan Ruediger, and Makala Skinner. 2022. "Streaming Media Licensing and Purchasing Practices at Academic Libraries: Survey Results." *Survey*. Ithaka S+R. https://doi.org/10.18665/sr.316793.

Cooper, Danielle and Rodger Schonfeld. 2021. "A Media Empire for the (Public) Library? OverDrive Acquires Kanopy." *The Scholarly Kitchen*. June 15, 2021. https://scholarlykitchen.sspnet.org/2021/06/15/overdrive-kanopy/.

farrelly, deg. 2016. "Digital Video: Merrily Merrily, Merrily, Merrily, Wading into the Stream: Freed from the Limitations of a Physical Artifact, Streaming Videos Are Available Anywhere, Anytime. No Need to Reserve in Advance, Arrange for Equipment, or Pick up and Return." *Computers in Libraries* 36 (9): 4–8.

Foley, Marianne. 2019. "Streaming Video PDA: Brace Yourself, Usage is Coming." *Proceedings of the Charleston Library Conference*. https://doi.org/10.5703/1288284317017.

Lohmann, Sam and Linda Frederiksen. 2018. "Faculty Awareness and Perception of Streaming Video for Teaching." *Collection Management* 43 (2): 101–19. https://doi.org/10.1080/01462679.2017.1382411.

Lowe, Randall A., Rick Davis, Leigh Ann DePope, Deborah K. Li, Robert K. Reeves, and Cindy Steinhoff. 2020. "Managing Streaming Video: The Experiences of Six Academic Libraries." *Journal of Electronic Resources Librarianship* 32 (2): 119–26. http:// doi.org/10.1080/1941126X.2020.1739831.

Netflix. n.d. "Educational Screenings of Documentaries." Accessed March 3, 2023. https://help.netflix.com/en/node/57695.

Polanka, Sue and Emilie Delquié. 2011. "Patron-Driven Business Models: History, Today's Landscape, and Opportunities." In David A. Swords (ed.), *Patron-Driven Acquisitions*, 119–36. Berlin, Boston: De Gruyter Saur.

Serrano, Amauri and Michael Fernandez. 2023. "Impacts of COVID-19: Toward a Streaming-Preferred Video Collection Policy." *Collection Management* 48 (1): 5–21. https://doi-org./10.1080/01462679.2022.2124392.

Spratt, Stephanie J., Gabrielle Wiersma, Rhonda Glazier, and Denise Pan. 2017. "Exploring the Evidence in Evidence-Based Acquisition." *The Serials Librarian* 72 (1–4): 183–9. https://doi.org/10.1080/0361526X.2017.1321901.

Urban, Sandra Gall. 2022. “Evaluating Kanopy Access Models in Academic Libraries: Balancing Demand and Budget Constraints.” *Collection Management* 47 (1): 37–48. https://doi-org./10.1080/01462679.2021.1940410.

Wang, Jian and Elsa Loftis. 2020. “The Library has Infinite Streaming Content, but are Users Infinitely Content? The Library Catalog vs. Vendor Platform Discovery.” *Journal of Electronic Resources* Librarianship 32 (2): 71–86. https://doi-org./10.1080/1941126X.2020.1739818.

3 Budgeting for Streaming Video

Introduction

The increase in user demand for streaming video for educational, research, and entertainment purposes has put a strain on already limited library collections budgets, many of which are shrinking or not keeping pace with inflation. Furthermore, the high cost and the number and fluidity of acquisitions models have made it difficult for libraries to adequately and sustainably fund streaming video. However, budgeting for video has always been complex and challenging due to the evolving nature of the format. Writing about the transition from VHS to DVD in 2004, Mary Laskowski noted that although new video technology can improve audio and video quality, it can also "play havoc with established collections and budgets" (Laskowski 2004, 217). Although Laskowski was referring to physical media, several of the collection development and budgetary challenges she highlights are broadly applicable to the current streaming landscape, including the variety of formats or purchase options available at any given time, lack of purchase options, repurchasing of owned content in new formats, and the need for new technology to play or view content.

This chapter outlines the budgetary challenges associated with streaming video and explores the process of budget planning and management of the library's collections budget from defining collection goals and objectives to the allocation and management funds. It provides negotiation and cost-containment strategies, discusses the use of assessment for effective budgeting, and provides practical budgeting strategies by acquisition model. The chapter focuses on academic libraries but also addresses acquisition models and collecting practices found in public libraries.

Budgetary Challenges

Although budgeting for collections is library-specific, survey data and the professional literature indicate that most libraries face similar budgetary challenges related to streaming video, including balancing user demand with high costs and budget constraints, navigating evolving acquisition and pricing models, and allocating and managing funds.

Recent surveys of academic libraries have found that libraries cite cost as the biggest challenge in acquiring and licensing streaming video, which is consistent with librarian interviews conducted for this publication and captured in the library vignettes at the end of each chapter (Tanasse 2021; Cooper, Ruediger, and Skinner 2022). In a 2020 survey, Frederick and Wolff-Eisenberg found that academic library directors expected to increase the amount allocated to streaming media from 5 percent in 2020 to 8 percent by 2025, making it the third largest allocation after e-journals/databases and e-books (Frederick and Wolff-Eisenberg 2020, 21). Two years later, Cooper, Ruediger, and Skinner found that academic libraries are not only spending a greater proportion of their collections budget on streaming video, but some respondents anticipate the amount allocated to double in five years (Cooper, Ruediger, and Skinner 2022). The picture is similar in public libraries as evidenced in *Library Journal*'s 2022 Public Library Materials Survey, in which the percentage allocated to streaming video in public libraries doubled from 2019 to 2021, from 3 to 6 percent, while during the same period, the percentage allocated to physical video dropped from 12 to 9 percent of the collections budget (Wyatt 2022, 28–9). To accommodate the increase in streaming video expenditures, libraries will likely need to find additional funding, repurpose funds from other sources (e.g., physical video, book approval plans), cancel continuing resources (e.g., databases, journal subscriptions, standing orders), or a combination of all the above. To stay within budget, libraries must clearly define collection priorities to focus on acquiring content that is most needed by their users.

The number and fluidity of acquisition and pricing models coupled with the lack of institutional streaming options for some content such as feature films and television series, adds another layer of complexity and unpredictability. As a result, budgetary strategies must be regularly revisited to ensure that the library is maximizing its spending power while also meeting local needs. Because of high cost and limited acquisition options, libraries often find themselves using a variety of models to meet current needs while balancing stable and long-term access to streaming video collections. Pricing models to lease and purchase streaming video further complicate budgeting. Cost is based on acquisition model and varies by distributor or streaming provider. As farrelly noted in 2014, there are no industry standards for streaming video pricing (farrelly 2014, 221), but in the last ten years, in part due to mergers and acquisitions and consortial agreements, it is possible to get a general sense of price by acquisition model, especially for use-driven models (i.e., PDA, EBA, PPU), term licenses, and perpetually purchased single titles which are often listed on content provider websites. Moreover, as libraries license more streaming video content, pricing patterns will emerge that will help negotiate reasonable pricing for new content, better estimate costs, and forecast streaming video allocations. Table 3.1 provides an overview of budgetary challenges by acquisition model and average pricing.

Table 3.1 Streaming Video: Pricing Models

Acquisition Model	Budgetary Challenge	Pricing Model and Cost
Patron-driven acquisitions	Expensive Difficult to control cost (unpredictable)	Annual license—approximately $150 to $250 per film Cost per video can vary by distributor/creator, genre, subject matter, etc.
Evidence-based acquisitions	Upfront financial commitment (possibly large) Difficult to sustain with flat or fluctuating budgets	Amount of content available is proportional to the upfront financial commitment Cost per title can vary by distributor/creator, genre, subject matter, etc.
Pay-per-use/Cost-per-circ	High cost per view	$2 to $6 per view Some providers cost-per-circ based on the length of the video and/or viewing period
Subscription-based	Expensive, especially for larger collections (e.g., Academic Video Online or Films on Demand) Annual inflationary increases Difficult to sustain with flat or fluctuating budgets	The fee may be based on the number of videos in subscription, FTE, tiered pricing by Carnegie classification
Perpetual	Expensive May require hosting fees if on streaming provider platform Indirect cost to locally host video (e.g., staff time and infrastructure)	Perpetual access license (vendor-hosted) Individual film—$300 to $500 Collection—depends on number of videos in collection Digital Site License (DSL) (locally hosting) Individual film—$200 to $1,000

The process of allocating funds and the responsibility for expending funds differs from library to library. For those libraries with complex budget structures and decentralized selection and fund management responsibilities, budgeting for streaming video and tracking expenditures is a challenge. Although *The ACRL Guidelines for Media Resources in Academic Libraries* recommends allocating funding directly to media resources (ACRL 2018), many academic libraries do not have a specific fund line for streaming media (Tanasse 2021; Dixon 2017; farrelly and Hutchinson 2014), making it difficult to strategically allocate funds and monitor expenditures. The process is more difficult when streaming video is funded on multiple fund lines, each managed by individual librarians who select streaming content, which often occurs in libraries that are building media collections but do not have media libraries or librarians. In this case, having clear and defined selection and fund management responsibilities is key to adequately fund streaming, track expenditures, and ensure proper stewardship of the budget.

Collection Development Philosophy and Goals

The *ACRL Guidelines for Media Resources in Academic Libraries* state that academic libraries should "provide adequate funding for media services and collections." "Adequate funding" is determined by the library's goals and objectives for media services and collections (Association of College and Research Libraries 2018). The objective of budget planning is to allocate funds to meet the needs of users and ensure that the collection aligns with the mission of the institution. The first step in the planning process is to set clear collection goals and priorities. Regardless of size, all academic libraries aim to meet the teaching, learning, and research needs of their user communities, but their collection development strategies may differ based on the institution's level of research activity, size of academic programs, and general collection development philosophy. Some may focus on meeting the immediate needs of their users while others may also seek to build collections to meet long-term teaching and research needs, as well as preserve the scholarly record. Like academic libraries, key elements of public library mission statements address the education and learning needs of their communities, but they also highlight culture, recreation, and community building (Fraser-Arnott 2022). These areas of focus shape the content libraries acquire and how they develop and provide access to information resources—in other words, the way in which they allocate funding and the amount they allocate for resources, including streaming video.

A library's collection development philosophy or approaches to selection will inform decision-making, which in turn has implications on budgeting. These approaches are regularly referred to as just-in-time and just-in-case, and Gary Handman adds a third, one that focuses on "selecting materials to meet the broad entertainment and educational needs of community users," which he describes as a circulating collection model (Handman 2010, 325–6). A just-in-case approach is a traditional collection development method, primarily found in academic libraries—especially at large research institutions—that focuses on building broad collections based on perceived user needs. They are resources that are deemed of value to local users and the general scholarly community. This approach requires a significant financial commitment on the part of the library, as well as the internal expertise to select materials. A just-in-time approach, on the other hand, looks to meet known user needs at the point of need. In the case of streaming video, it is often used to meet specific curricular, research, or programming needs such as course reserves, user requests, and campus film screenings. The circulating collection approach, which focuses on recreational and educational needs, is usually found in public libraries whose collections often consist of current and popular materials many of which may not be available streaming for institutional use.

Most academic libraries will use a combination of all three approaches based on specific use cases and available funding. In the case of streaming video, libraries have tended to take a just-in-time approach in part due to the high cost of licensing streaming compared to physical video equivalents and the lack of perpetual purchase options. For many libraries guaranteed usage informs how they license streaming video, and therefore, many libraries pursue use-driven models (i.e., PDA, EBA, and pay-per-use), in which titles are licensed based on patron discovery and use rather than the anticipation of patron use. Libraries that prioritize collection building, and have the budgetary means to do so, are more likely to proactively license video, and perpetually purchase titles when available to expand and strengthen the library's collection. However, the availability of funds can fluctuate throughout the fiscal year and specific use cases may change the library's general collection development strategy. For example, a library that normally takes a just-in-time approach may proactively perpetually license a streaming video collection if end-of-year or one-time funds are available.

Although the various collection development approaches inform the library's preference for owning and leasing content depending on use case and available acquisition models, the collections budget is finite, and setting clear collecting priorities is important to adequately and effectively budget for video. Priorities can be based on use (e.g., curricular or recreational), format (e.g., physical or streaming video), or streaming provider (e.g., Hoopla or Kanopy). For academic libraries of all types, instruction is often the most important factor for licensing streaming video (Cooper, Ruediger, and Skinner 2022, 9); therefore, filling course reserve requests will likely be a higher priority than filling individual streaming video research requests. For institutions with film and media studies programs, libraries may choose to prioritize collection building of hybrid physical and streaming collections in which streaming provides greater access to content, but physical copies are acquired when streaming is not available for perpetual purchase (or at all), or when the physical copy provides better picture quality, is a different version, includes commentaries, or is better for close analysis. Public libraries may prioritize providing access to streaming video via platforms that also offer other material types such as audiobooks and e-books to maximize funding (e.g., Hoopla).

Budgeting Process

Budget Basics

Once collection goals and priorities are defined, budget planning begins for the next fiscal year, but before delving into budgeting for streaming video, a general understanding of the collections budget is useful. The collections budget is essentially a planning tool with a financial allocation that "serves to document selection decisions through allocations" and track progress to meet collection goals (Johnson 2018, 83). When discussing the collections budget, it is useful to think of it in three parts: the financial allocation (i.e., the budget), the process of allocating funds (budgeting), and the management of those funds (budget management) (Burnette 2008, 4). The collections budget normally covers the cost to acquire information resources but may also cover costs for collections-related services such as processing (e.g., shipping), cataloging (e.g., MARC records), access and hosting fees, discovery of collections, collection analysis tools, open-access initiatives, third-party repositories, and sometimes, consortial memberships. The scope of the collections budget will not only affect the financial allocation but also the plan to spend funds and the complexity of managing them.

In addition to the budgetary challenges specifically associated with streaming video, budgeting for collections in general must account for a variety of factors beyond collection development goals and priorities, or the scope of the collections budget. These factors include the fund structure; selection and fund management responsibilities; the amount of non-discretionary (i.e., amount of the budget committed for continuing resources) and discretionary funds; funding sources (e.g., town, state, university) and their availability and reliability; pricing models and trends; inflationary increases; trends in usage and demand; the fluidity of economic cycles; and institutional accounting requirements. However, the main goal is to maximize the budget to meet the library's short- and long-term collection development objectives.

There are different approaches to budgeting for collections. Some libraries use allocation formulas that use weighted variables to calculate allocations such as student enrollment, cost of materials, use, number of faculty, programs, and library and university goals. However, the easiest approach to budgeting, and the one most commonly used in libraries, is baseline budgeting in which the library begins with the previous year's collections budget allocation and expenditures as a starting point, and fund lines are increased or decreased proportionally based on available institutional funding and/or endowment yield for the next fiscal year. The budgeting strategies presented here use a baseline budgeting approach.

Budget Forecasting

The amount allocated to streaming video will depend on the library's collection development objectives and priorities and the size and stability of the collections budget. Estimating costs to license streaming video as accurately as possible is key to good budget planning. The easiest place to start is with the previous year's streaming video allocation(s) and expenditures which is essentially a review of the previous year's plan to spend the funds and to what extent the library successfully estimated costs and anticipated curricular and research needs. It is an opportunity to evaluate the effectiveness of the library's current budget strategy and make adjustments if needed. Budgeting for streaming media requires a certain amount of trial and error to determine appropriate acquisition models, a reasonable budget, and how to best allocate funds and track expenditures. Furthermore, providers regularly release new content and add or tweak their acquisition and pricing models all of which affect budget planning and cost estimates.

Recommendations for budget planning and forecasting:

- Gather information to inform cost estimates: negotiated inflationary increases and discounts, expenditure data for course reserves, user requests, and historical expenditure date for "collection building".
- Work with content providers to explore new purchase or pricing models that better suit the library's needs and request estimates when applicable.
- Assess use and cost for ongoing commitments (e.g., subscriptions) to inform renewal decisions.
- Review historical data for short-term licenses and identify potential perpetual purchases.
- Review budget requests for new streaming content (e.g., subject librarian's request for a new drama collection) and identify those that align with selection criteria and collection development priorities.
- Investigate consortial offers for currently subscribed collections, regularly purchased content, or new streaming content of potential interest.
- Consider streaming pilots for new resources or acquisition models to determine if beneficial; pilots provide cost and usage data which help estimate costs.
- Explore reallocating funds to sustain current streaming expenditures and potential new acquisitions (e.g., reallocate physical media or print book funds for streaming video).
- Develop a desiderata list for potential one-time purchases, if one-time funding is available (e.g., grants and gifts).
- Have a contingency plan for expensive requests or unexpected increases.

Funding Sources

Funding sources are an important factor in determining how funds are expended. Funding sources vary by library type, are susceptible to fluctuations in the economy, and may not be permanent sources of revenue. Public libraries are funded primarily with local or state property taxes, and to a lesser extent with federal funding, private donations (e.g., friends of the library), government and corporate grants, or endowments. Since most funding is tied to taxes, there are often significant funding disparities between urban and suburban libraries and rural libraries, thus affecting what libraries acquire and how much they can acquire. Academic libraries, on the other hand, are usually funded by their parent institution, endowments, grants, and gifts, and may also receive supplemental funding from the state or county, particularly in the case of public universities and community colleges. Endowments and grants usually stipulate how funds must be used, which can be broad or narrow in scope, such as funds to purchase content with a diversity, equity, and inclusion (DEI) focus versus content about underrepresented groups in urban areas. Grants are usually one-time funds available for a defined period, such as ALA's COVID Library Relief Fund, which provided emergency funding in 2021 to tribal, school, public, and academic libraries experiencing hardships due to the COVID-19 pandemic (American Library Association 2021). It is not good practice to use one-time funds on continuing resources (e.g., subscriptions) unless the library plans to absorb the cost in the next fiscal year's collections budget. For example, a public library may choose to use town and state funds for their Hoopla pay per use plan, and grants for one-time purchases and DVDs. Whereas an academic library may choose to use sustained funding for streaming video subscriptions, yields on endowments with narrow gift agreements for subject or thematic streaming collections, and one-time funding for streaming pilots or perpetual purchases. Funding sources are an important component in determining how funds are allocated and the amount allocated. However, in some colleges and universities, funding for streaming may also come from other campus units, and the library may have little to no control over the availability of funds or how they are spent. For example, at Delaware County Community College in Media, Pennsylvania, the library manages the college's streaming video collections and makes renewal recommendations, but the Office of Online Learning and the VP of Finance and Administration negotiate pricing, license, pay, and make final renewal decisions for streaming video collections (Rodgers 2023).

Allocating and Managing Funds

The collections budget structure differs from library to library, but in general, most libraries use a line-item format to allocate funds because it is simple and relies on historical expenditure data to estimate allocations. In a line-item budget, funds are allocated by category which can be broad or narrow in scope. Fund lines can be based on subject, material type and format (e.g., video or e-video), collection (e.g., graphic novels), department, program, user group, or purchase model (e.g., recurring, one-time, approval plan, use-driven), or specific grants and endowments. The fund structure itself can vary in complexity and may reflect types of expenditures (e.g., discretionary or non-discretionary), historical practice, the library's organizational structure (e.g., subject librarians or collection managers), revenue sources, and the size and scope of the library's collection. The fund structure will dictate how funds are allocated, the responsibility to expend funds, and the ease with which expenditures are tracked. The allocation process can be as simple as a public library that allocates $7,000 per year for a Kanopy PPU plan funded with the e-resources fund line that is managed by the library director, and if use exceeds expected demand, the number of allowed video checkouts is reduced. On the opposite end of the spectrum, at Yale University Library, streaming is integrated in to the library's budget structure like other e-resources. Multi-subject and interdisciplinary streaming video subscriptions, mediated PDA plans, and multi-subject one-time streaming purchases are funded centrally and managed by the Central Collection Strategist. Course reserves are funded with a designated course reserve fund, while subject funds are used for requests and collection building based on subject-specific criteria (Yale University Library 2023).

The ACRL Guidelines for Media Resources recommends allocating funds specifically for media as part of the collections budget. However, survey data from 2015 to 2021 has consistently shown that only about a third of academic libraries have a defined budget line for streaming video (farrelly and Hutchison 2014; Dixon 2017; Tanasse 2021). However, a 2022 survey of 228 academic libraries conducted by Austin and Thorton, found that two-thirds of respondents have a "planned budget line item" for streaming video (Austin and Thornton 2022, 248). The large increase in the Austin and Thornton survey may be due to the fact that it included more libraries and the majority of respondents were at institutions with a FTE below 5,000—that is, smaller institutions with smaller budgets and simpler budget structures. For budgeting purposes, a media or streaming video line item makes it easier to allocate and manage funds, make adjustments as needed, and simplifies decision-making and the renewal process. However, a line item for video may not be a feasible option for libraries with no media library or individual responsible for managing the media/streaming video budget. This is particularly true for libraries whose budget structure has fund lines for subjects, acquisition models (e.g., ongoing, approval, PDA), and endowments with multiple librarians and groups selecting video content and managing individual funds. Regardless of their budget structure, libraries should aim to create what Fleming refers to as a "data-rich environment." This involves using tools available in the library services platform (LSP) to code and "develop cascading levels of detail for budget tracking and assessment" (Fleming 2015, 175). A well-organized collections budget includes "facets," such as material type (e.g., video, e-video), attached to individual subjects, course reserves, and general e-resources funds. This ensures that allocation and expenditure data are collected, enabling efficient budgeting for streaming video and tracking expenditures over time.

Assessment and Budgeting

Although the assessment of streaming video is addressed in-depth in Chapter 9, a brief overview of the use of quantitative data, in particular financial metrics, in collection development decision-making is useful as it has budgetary implications. Quantitative data can be used to assess the success of a streaming pilot, make renewal decisions, reallocate funds, or advocate for sustained or additional funding. Financial metrics such as cost-per-use (CPU), cost per title, return on investment (ROI), and cost–benefit analysis (CBA) are useful analysis tools to help demonstrate the value of streaming video content.

CPU and cost per title are monetary metrics that can be easily calculated and are useful for making renewal decisions, renegotiating subscription fees and inflationary increases, and choosing an appropriate acquisition model for the library's needs. For example, a library may choose not to renew a subscription for a large streaming collection if the CPU (e.g., subscription fee divided by the total number of views during the subscription period) is high compared to a similar library-subscribed collection, there is significant overlap with other licensed collections, the annual increase was high by library standards, or there is not enough unique course reserve content to justify the subscription. However, if the subscription includes access to exclusive content that is used for curricular and research purposes, the library may choose to use data gathered (e.g., cost-per-use/title, amount of duplication) as leverage to renegotiate the subscription fee or lower the inflationary increase.

ROI and CBA are financial analysis measures that can be used to determine how to best allocate resources, especially when choosing among alternative acquisition models, and to advocate for additional or sustained funding for streaming in general. A 2010 article by Pan and Fong describes the use of ROI and CBA to demonstrate the value of collaboratively purchased e-resources at the University of Colorado Libraries and provides useful and clear definitions and formulas of ROI and CBA that are applicable to streaming video. ROI is expressed as a percentage that shows the return or increase on dollars spent to achieve a benefit ((BENEFIT − COST) ÷ COST) x 100). CBA is expressed as a ratio showing the dollar value of benefits gained for dollar values of costs (BENEFIT ÷ COST) (Pan and Fong 2010, 190). ROI and CBA are useful metrics to compare and evaluate alternate streaming acquisition models as evidenced in a 2017 pilot also conducted at the University of Colorado system, that compared the cost and content of consortial EBA and PDA streaming plans and used ROI and CBA metrics to help evaluate the success of the pilot (Spratt et al. 2017).

Although financial metrics are useful indicators of value, they do not provide a complete picture of value when assessing the impact of library-provided streaming video content on teaching and learning. Financial metrics as defined above rely on usage data to quantify value; however, measuring and interpreting video use is complex, not only because some content providers do not provide COUNTER-compliant data making it difficult to compare usage across resources, but the interactive nature of video adds another layer of complexity. Financial metrics can enhance decision-making, but it is important to consider how usage is measured when calculating and interpreting the results. In a 2015 issue of *Against the Grain*,

David Parker, head of the editorial and licensing department at Alexander Street, questioned the use of playbacks to illustrate the value of PDA, highlighting that use can be defined by the "length of views (engagement), number of classes where a particular video is central to the syllabus (impact of view), or longevity of views (number of years the video remains active within the academic community)" (Parker 2015, 72). He recommended that libraries take a more nuanced approach to measuring value beyond ROI. Since academic libraries cite curricular support as the primary driver for streaming video acquisitions, demonstrating impact on students is key. A possible approach is to gather expenditure data for streaming video course reserves (when possible) and tie reserve requests to specific courses to determine the number of students in each course. Using this data, the library could calculate streaming video cost per student thus demonstrating the impact of streaming video expenditures on teaching. Although gathering this data is easier said than done, it better informs the library's collection development strategy and makes a stronger case for additional funding.

Negotiation and Cost-Containment Strategies

In the current climate of rising prices, ever-increasing numbers of e-resources, and diminishing library budgets, negotiation is essential to effectively steward the library's limited budget. Librarian negotiation skills are a critical component to ensure that the library maximizes its purchasing power. Flowers describes effective negotiations as those that "are beneficial to all concerned and serve as the vehicle for meeting the ultimate goals of both parties [i.e., vendor and library]" (Flowers 2004, 435). Librarians should prepare for the negotiation process to effectively communicate with content providers and propose possible solutions that benefit the library and the provider. It is a useful strategy for librarians to view the negotiation from the provider's perspective to identify ways in which the proposed solutions can benefit the content provider's bottom line and simplify their work. For example, a library is more likely to pursue streaming video perpetual access licenses if hosting fees are waived, and if fees are waived, the vendor has the benefit of not tracking hosting fees and issuing annual invoices. The following tips will help facilitate the negotiation process and increase the likelihood of favorable pricing.

Preparation Tips:

- Know the needs of your user community.
- Determine if currently licensed streaming video content is meeting the needs of users.
- Review currently licensed streaming video content and similar products and do a CBA.
- Determine how much funding is available.
- Know the streaming provider and their acquisition models.

- Review available acquisition models and consider strategies to negotiate pricing with the understanding that the amount of room for negotiation varies by model.
- For renewals, review usage and expenditure data.

Cost-Containment Strategies and Implications

- *Negotiation of annual inflationary increases and hosting fees when licensing content for the first time or at the point of renewal*
  Depending on the cost of the subscription, annual inflationary increases can be substantial and have a negative impact on the collections budget, especially for budgets that are declining, flat, or not increasing at the rate of inflation. Furthermore, they can limit the amount of new content the library can license and impact renewal decisions for other resources. Subscription fees for larger collections can be upwards of $25,000, and a standard 5 percent annual increase translates into an additional $1,250 which is then added to next year's subscription fee. Renewals offer the opportunity for savings. It is good practice to compare the costs of similarly licensed resources and request a reduction in cost if the product in question is more expensive. For perpetually purchased content, a streaming provider may charge an annual platform or hosting fee per video which should be negotiated at the time of purchase. Possible options could include the waiver of hosting fees if the library maintains an agreed-upon spend, delay in paying hosting fees until an agreed-upon date (e.g., three years after purchase), or a flat platform fee for all streaming content purchased and hosted on the platform.
- *Consortially negotiated pricing and discounts or access to streaming content as part of consortial membership*
  State and national consortia use their collective influence and greater purchasing power to get more content at a reduced cost. For streaming video, this could be a percent discount on subscription fees or one-time purchases, negotiated inflationary increases, or negotiated tiered pricing by FTE or user population. Some state consortia license and provide access to streaming video collections for their members such as VIVA (Virtual Library of Virginia) and NC Live.
- *Multi-year agreements and annual increase caps (e.g., subscription-based collections)*
  For libraries whose institutions allow multi-year agreements and are confident that the collections budget is stable and can absorb the annual fees, multi-year agreements are a good approach to lock in favorable pricing and annual increases. Multi-year agreements are often three to five years and can also facilitate budget planning and forecasting. However, they can also limit the library's flexibility and ability to repurpose those funds for unanticipated needs such as requests for new resources, an invoice that was mistakenly not paid the previous year, or an unexpected subscription increase.

- *Leverage total vendor acquisitions for better pricing*
  Some vendors offer streaming content and other e-resources (e.g., e-books, article databases) and libraries can leverage annual spending with the vendor to get a better discount on the streaming product in question (e.g., Alexander Street (ProQuest/Clarivate) and Films on Demand (Infobase)).
- *Bulk and bundled discounts and pricing*
  Streaming providers will often offer a discount for large purchases, especially one-time purchases of streaming video collections. Some streaming providers and distributors will offer reduced pricing for bundles of films purchased for life of file or leased for a defined period. Independent distributors are often open to discounts if a library acquires multiple digital site licenses (DSL) for life of file at one time.
- *End-of-fiscal or calendar-year discounts, especially useful for one-time funds*
  Streaming providers and independent distributors usually offer end-of-year discounts at the end of calendar year or their fiscal year. It is generally good practice to know when a vendor's fiscal year starts and ends and to plan accordingly.
- *Deposit accounts*
  If institutional accounting rules allow, deposit accounts may be a good option when one-time funds are available that must be spent in the current fiscal year. The deposit account does not reduce cost per se, but it allows the library to spend the funds across fiscal years. In other words, the deposit account frees up funds in the new fiscal year and helps mitigate potential expensive requests or price increases.
- *Mediate and control user requests or checkouts*
  For patron-driven acquisitions, libraries can limit the number of course reserve requests per class or user, as well as define criteria to approve requests (e.g., curricular and research purposes). For a PPU model, libraries can limit the number of loans per user. A 2022 survey of academic libraries that use Kanopy PDA models (unmediated, mediated, Smart PDA, and others) found that some libraries use different models based on the time of year such as "PDA during the semester and mediated access during breaks" thus limiting staff time to mediate likely curricular requests during the semester while mediating likely recreational use during the breaks (Urban 2022, 43). Another approach to mediation is to refer users to the library's DVD collection when expected use does not justify the cost of a streaming license or when the request falls outside of the library's collecting priorities (e.g., request for entertainment purposes in an academic library).
- *Discount for content owned on DVD*
  In the case of in-perpetuity licenses that are locally managed, many educational distributors will offer a discount on streaming licenses if the library currently owns the film on DVD and the DVD was purchased from the distributor.

Budgeting by Acquisition Model

Although we can categorize streaming video acquisition models into groups (in-perpetuity models, term models, subscriptions, use-driven—PDA, EBA, PPU), each streaming provider has their own flavor of the model which will have budgetary and workflow implications requiring different strategies to allocate funds, expend those funds, and track expenditures. Therefore, it is important to work with content providers—they know their products and can use their experience with other libraries to propose solutions and strategies for local needs. Furthermore, budgeting in general requires a certain level of creativity and flexibility, but budgeting for streaming video, especially for specific streaming provider models, can benefit from input and advice from colleagues who are dealing with the same issues. Logically, the library literature is a useful source, but the VideoLib list (videolib@lists.berkeley.edu) is an excellent resource for getting real-time advice on specific concerns and questions.

Use-Driven Acquisition Models

Although use-driven acquisitions (UDA) models such as PDA, EBA, and PPU provide users with a large and diverse collection of videos and reduce staff time to both select and process content, they also present a series of budgetary and associated workflow challenges that should be considered before entering into a UDA program. Developing a budgetary strategy to sustainably support use-driven models is instrumental for the long-term success of the program. According to Carrico, Leonard, and Gallagher, libraries must often find creative ways to fund UDA, and they propose a couple of budget scenarios that are applicable to streaming video: "one-time funding for pilots or short-term projects and annual funding for sustained programs" (Carrico, Leonard, and Gallagher 2019, 17). It is useful to project costs with the use of historical expenditure data; therefore, pilots offer libraries new to UDA the opportunity to gauge spending and refine the program as needed to stay within budget.

Use-Driven: Patron-Driven Acquisitions

Patron-driven acquisitions allows the user to make acquisition decisions at the point of need by triggering a short-term license or via a request for mediated PDA plans. They can be unmediated, mediated, or curated, and each of these options informs how the library budgets and the amount of control it has over spend. As evidenced in the library literature, it is difficult to control costs for unmediated PDA programs which are often unsustainable. In fact, many libraries have opted to suspend unmediated PDA programs and move to mediated models (Foley 2018; Wang and Loftis 2020; Urban 2022) or curated collections (Prelitz 2022) for which they have more control and can better predict spend. Libraries implement streaming PDA programs mainly for two reasons: personal edification (public libraries) and curricular and research support (academic libraries). The amount allocated to PDA programs is dependent on the intended purpose of the PDA collection and the amount of funds available. If the intended use is curricular support, the library should consider budgeting for the collection like other course reserve materials, and use historical expenditure data to estimate the next fiscal year's allocation.

It is beneficial to work with streaming providers to curate the collection to ensure that triggered videos align with the library's collection goals and available funding. After launching a Kanopy PDA collection in 2018, Delaware County Community College (DCCC) found that they had expended their entire budget for the program four months into the fiscal year due to the unexpected use of *The Great Courses* series as lecture material. The PDA collection was intended to provide access to films to support students' coursework and research needs and not serve as lecture material. Fortunately, DCCC librarians were able to secure additional funding and remove the series from the PDA pool, and although viewing usage remained high, the library was able to stay within budget (Rodgers 2018).

Deposit accounts are useful to track expenditures and stay within budget, and to strategically use one-time funds across fiscal years. For example, a library can turn off an unmediated PDA plan if it is quickly depleting the allocated budget and move to a mediated model or a library may opt to place end-of-year funds (i.e., use it or lose it funds) in a deposit account so that funds can be used the following fiscal year. The use of blanket POs for PDA programs is another approach to track expenditures for those libraries whose institutional accounting requirements do not allow deposit accounts. Depending on the role PDA plays within the library's overall collection development strategy, pilots are useful to gauge use for new programs and gather expenditure data to determine fiscal year allocations.

In addition, usage data can help identify potential candidates for perpetual purchase which may be available from the same content provider (e.g., Kanopy, Alexander Street, and Infobase) or directly from educational distributors. Furthermore, many PDA providers offer subscriptions to content on the platform, and usage data can help determine if it is more cost-effective to pursue a subscription rather than individual short-term licenses. For example, if the Kanopy BASE subscription is equal to or less than the cost of triggered PDA licenses included in the Kanopy BASE subscription, the library may choose to subscribe and get access to more content while continuing to use PDA to supplement the subscription.

Use-Driven: Evidence-Based Acquisitions

In an EBA model, the library pays an upfront fee for access to a defined number of streaming titles for a specified period, usually one year. The financial commitment is proportional to the number of titles included in the collection. At the end of the program, the library selects titles for perpetual purchase based on usage statistics and locally developed selection criteria. A pilot project at the University of Colorado

Libraries, comparing EBA and PDA, found that EBA may be a better choice for libraries with the financial resources to perpetually purchase rather than pursue short-term licenses (Spratt et al. 2017). Budgeting for EBA is similar to budgeting for a subscription. Some providers market their EBA programs as subscriptions with the added benefit of perpetual ownership of a select number of videos in the collection, such as Alexander Street's AVON Build by Choice program. Like databases, the library can choose to treat EBA programs as ongoing commitments, thus committing funds for subsequent fiscal years. One-time funds can be used for new EBA pilots to determine if usage warrants the financial commitment, as well as providing the data necessary to curate the collection.

Use-Driven: Pay-Per-Use

The PPU model is commonly found in public libraries, in which the library pays a fee each time a video is "checked out" for viewing for a defined period, usually seventy-two hours. For public libraries, the cost to provide access to streaming video can be prohibitive in light of fluctuating budgets, which are primarily based on town property or sales taxes, as well as the size of the population served. However, a survey conducted by Kanopy in fall 2020 found that public libraries expected their streaming budgets to grow, primarily at the expense of physical media acquisitions. Furthermore, most respondents noted that there was an increase in streaming video usage during the COVID-19 pandemic, and they expected usage to remain steady (Kanopy 2020). Ed Garcia, Director of Cranston Public Library in Rhode Island, in an interview for this publication, noted that streaming video usage at the library increased during the pandemic but use has plateaued. In fact, there are no plans to increase funding for the library's Hoopla and Kanopy streaming video services and the library will continue to acquire DVDs and Blu-rays which still regularly circulate (Garcia 2023).

Short-Term Model

Term models are defined here as single titles or multi-part series usually licensed for one to three years. These titles are usually requested for teaching, research, or institutional programming (e.g., Women's History Month documentary) purposes. Since these titles are requested, it is difficult to estimate user requests that would fall under this category and the amount to allocate since pricing models can differ greatly by provider. Some providers may be new to working with libraries thus pricing may be unpredictable. For short-term licenses, most of which are used for course reserves, libraries should budget like they do for course reserves, preferably using historical streaming expenditures for short-term licenses plus an inflationary increase of 5 percent or the increase that the library uses for annual budget modeling. If funds are needed beyond this amount, and considering that curricular support is likely the primary reason for licensing streaming video, the use of discretionary funds for proactive collection building or contingency funding can be used to supplement the course reserve budget. Depending on the budget structure, the funds can be allocated to fund lines for course reserves, video, streaming video, e-resources, etc. but the important thing is the ability to track expenditures to better forecast cost and to determine if additional funds are needed for the new fiscal year.

Subscription Model

Streaming video subscription packages offer access to a large volume of content, predictable costs, ease of management, and consortial discounts are often available. Recent surveys of academic libraries have consistently found that the subscription model is the most popular and widely adopted model (Tanasse 2021; Cooper, Ruediger, and Skinner 2022; Austin and Thornton 2022). Subscriptions vary in cost, amount of content included, subject matter (multi-subject or discipline- or subject-specific), and genre (feature films or documentaries). All these factors influence the library's decision to subscribe and how it budgets. Streaming subscriptions can be handled like databases or e-content packages. In other words, they can be treated as ongoing commitments that are budgeted for and renewed per the library's collection development policy and practices. For example, large aggregated streaming video collections could be treated like large multi-subject publisher e-journal packages that are funded centrally whereas subject (e.g., Dance in Video from Alexander Street) or discipline-specific (e.g., Journal of Visualized Experiments (JoVE)) subscriptions can be managed and budgeted for at the subject level and included in annual subject budget requests like any other core e-resource. Although sustained funding for subscriptions is recommended, one-time funds for new subscriptions is an option to explore if use justifies the cost of the subscription.

Like e-content packages and databases, streaming subscription fees and annual inflationary increases are negotiable. Subscription decisions are based on the appropriateness of content for users and the quality and quantity of content. These factors—along with cost-per-use, amount of course reserve content included in the package, and amount of content added and removed during the subscription term—should inform renewal decisions. Even if the library is not considering cancelling the subscription, when applicable, these factors could help negotiate a lower subscription fee or annual increase.

Single Title (Perpetual Access: Digital Site License and Perpetual Access License)

Single-title perpetual licenses can range from $150 to $1,000 per title. DSLs have additional staff costs to receive files or digitize DVDs and upload files to a server or third-party service, as well as the cost to host video files (i.e., server or hosting costs). PALs may require the payment of annual hosting fees per film or a flat platform fee. Title-by-title perpetual purchases are meant to develop the collection and provide long-term access and can be treated like any other e-resource (e.g., e-books). For example, subject librarians or collection managers may budget for title-by-title streaming videos as they previously did for tangible videos with the understanding that streaming video is significantly more expensive; therefore, collection priorities should be well-defined. Limited staffing capacity may restrict the number of videos purchased that require local hosting. The library may also choose to identify videos regularly used for teaching and add the cost to acquire them in perpetuity to the new fiscal year's budget. Rather than pursue short-term licenses for course reserve requests, some libraries may choose to budget for perpetual access licenses for curricular support, especially if the subject content is a collecting strength or area of expanding focus. If discretionary funds are available and/or a discount is available, the library may choose to pursue PALs for titles that are being removed from a subscription. For libraries building video collections, it should be standard practice to earmark funding for perpetual purchases based on collection development priorities such as replacing heavily used titles on DVD with streaming.

Collections (Perpetual Access)

Perpetual purchase of streaming video collections is similar to the purchase of digital archives: they are expensive, often purchased with one-time funds, and trials are useful to analyze and review content for quality and appropriateness before making a purchase. If the content is core to a discipline, such as dance or drama, librarians may choose to include the cost of subject collections in their annual budget request. Librarians may keep desiderata lists and trial collections if one-time or end-of-year funding becomes available.

Conclusion

It is a challenge to provide general budgetary guidance for streaming video because local needs, requirements, and funding differ significantly from library to library. Budgeting is complex and must account for a variety of factors that are library-specific including the size of the collections budgets, collection development priorities, funding sources, staffing capacities, available infrastructure, fund structure and management, and ability to divert funding from other areas of the budget. However, one thing is certain—an appropriate budget strategy for streaming video must balance user demand and cost. It must be flexible and regularly reevaluated to accommodate new acquisition models, fluctuations in funding, and changes in user needs and institutional priorities.

Streaming Vignette—Chemeketa College

Name of the library: Chemeketa Community College Library

Reporting librarian and title: Jennifer Cox, Dean of Library and Learning Resources

Type of library: Community college, main library located in Salem, Oregon; and Yamhill Valley library, located in McMinnville, Oregon

User community: Approximately 17,900 students (headcount); more than half are part-time and nearly a third of students are enrolled exclusively in distance education courses

Total streaming budget: 2 percent to 3 percent of annual collections budget

Chemeketa Community College is the second-largest community college in the state of Oregon. It offers associate degrees, job training, General Education Development (GED) programs, and enrichment courses, and has recently started to offer bachelor's degrees. The library acquires materials in support of Chemeketa's instructional programs and houses the campus computer center. Jennifer Cox, the Dean of Library and Learning Resources, sees community college libraries as the bridge between public libraries and academic libraries at four-year and graduate degree-granting institutions. She highlights that community college library collections reflect the region in which they are located and fill gaps in the community. Although the college offers enrichment courses and certification programs (e.g., Cultural Competency Certificate), the library collection tends to cater to college students rather than public library patrons per se. In addition to the college library, community members have access to a wide range of resources and materials provided by public libraries in the area.

The library is not only reliant on its own funding sources to provide access to streaming video, but also uses its partnerships to expand streaming access for its users. It promotes partner resources and negotiates licenses and pricing via consortia and state associations, or campus units that support access and accessibility to streaming video. The library subscribes to Kanopy BASE, a package of more than 10,000 films from various distributors, which is complemented by a pay-per-use Kanopy collection provided by the Chemeketa Cooperative Regional Library Service (CCRLS). CCRLS is a cooperative of seventeen public libraries in the tri-county area and Chemeketa Community College Library. It is located on the Chemeketa Community College Salem campus and has close ties to the library, but the cooperative negotiates its own e-resource licenses, and its resources are currently only accessible to public library card holders. Although access is not ideal for Chemeketa Community College users, who must have a public library card, the library includes CCRLS resources in their A-to-Z database list and helps users sign up for public library cards. The library licenses AVON and Kanopy via the Orbis Cascade Alliance, an academic library consortium in the Pacific Northwest, and Films on Demand through the Oregon Community College Association's distance education subgroup. It also has smaller direct subscriptions to Infobase's Archival Films & Newsreels Video Collection and Alexander Street's PBS Video Collection. In addition to subscribed collections, the library also does title-by-title short-term licenses and purchases individual titles for life of file that are hosted on Kaltura, a media platform managed by the college's Center for Academic Innovation.

The library has a small staff of paraprofessionals, part-time librarians, three full-time faculty librarians, and the Dean of Library and Learning Resources. The selection of video, especially multidisciplinary and interdisciplinary content, is a collaborative process in which potential acquisitions and subscriptions are discussed as a group factoring in user needs and available funding. Prior to the pandemic, the library was actively purchasing DVDs, however, the decrease in foot traffic post-pandemic has resulted in a substantial decrease in the number of physical items purchased and a sharp increase in the acquisition of e-resources. Not only is the library acquiring fewer physical items, but it has also weeded some of the physical video collection to free up space to store laptops needed for the computer center. Changes in user behavior and expectations, as well as faculty requests, have shaped the library's collecting strategy. Currently, streaming is the preferred video format, and perpetual access is preferred with the understanding that faculty members will likely use the same films in the future. However, the library has recently begun to pursue more short-term licenses for films used in topic-focused courses, which is particularly true for film courses. In such cases, the library maintains a spreadsheet to keep track of short-term licenses and contacts faculty before licenses expire to determine if they need to be extended.

Accessibility, in particular closed captioning, is an important factor in the selection process and how streaming video is managed. According to Cox, accessibility is one of the biggest challenges in licensing streaming video. In fact, the library will not make purchased content available to users until it has captions. Cox believes that in the future the college will likely make captions mandatory for all video content created or purchased by the college. For locally hosted content, the library works with the Center for Academic Innovation and Accessibility Services to fund and create captions as well as a third-party provider to add captions to locally hosted videos.

The collections budget is not large, and Cox notes that it is therefore important to be good stewards of the library's funds. Streaming video is expensive, and the move from physical to digital has impacted fund allocations. During the pandemic, funds allocated for physical collections were repurposed for e-resources, a policy that has continued and will continue for the foreseeable future. The library has a small pool of endowed funds that are used for one-time purchases or e-resource pilots like the recent subscription to Met Opera on Demand. Usage statistics play an important role in managing the budget by informing selection decisions—such as in the case of trials, e-resource renewals, and the reevaluation of current collecting strategies. For example, usage statistics have shown that individually licensed videos get significantly more views than content in subscribed collections; therefore, it may be time to revisit available purchase models and content from new content providers.

No matter the type of library, Cox maintains that funding streaming video is expensive and challenging. It requires a budgetary strategy that involves collaboration and cooperation of library staff, campus units, and consortial and state partners. The library has leveraged these relationships to provide greater access to streaming video at a lower cost. Marketing these resources is also important to ensure that the library is getting a good return on investment. Library staff actively promote streaming video content to faculty and students via library instruction, workshops, and blog posts.

Chemeketa Community College Library Collection Development Guidelines: https://library.chemeketa.edu/about-the-library/library-policies/library-collection-development-guideline/

References

American Library Association. 2021. "The ALA Library Relief Fund." Accessed September 3, 2023. https://www.ala.org/aboutala/ala-covid-library-relief-fund.

Association of College and Research Libraries (ACRL). 2018. "Guidelines for Media Resources in Academic Libraries." https://www.ala.org/acrl/standards/mediaresources.

Austin, Cathy and Bonnie Thornton. 2022. "Streaming Video and Academic Libraries: A Survey on Commonalities in Acquisitions." *Journal of Electronic Resources Librarianship* 34 (3): 245–55. https://doi.org/10.1080/1941126X.2022.2099040.

Burnette, Elizabeth S. 2008. "Budgeting and Acquisitions." In Maria D. D. Collins and Patrick L. Carr (eds.), *Managing the Transition from Print to Electronic Journals and Resources: A Guide for Information Professionals*, 3–27. New York: Routledge.

Carrico, Steve, Michelle Leonard, and Erin Gallagher. 2019. *Implementing and Assessing Use-driven Acquisitions: A Practical Guide for Librarians*. Lanham, Maryland: Rowman & Littlefield.

Cooper, Danielle, Dylan Ruediger, and Makala Skinner. 2022. "Streaming Media Licensing and Purchasing Practices at Academic Libraries: Survey Results." *Survey*. Ithaka S+R. https://sr.ithaka.org/publications/streaming-media-licensing-and-purchasing-practices-at-academic-libraries/.

Dixon, Jennifer. 2017 "The Academic Mainstream: Streaming Video is Becoming a Mainstay on College Campuses, but Discovery and Walled-off Content Create Their Own Challenges." *Library Journal* 142 (14): 42–5.

farrelly, deg and Jane Hutchison. 2014. "ATG Special Report: Academic Library Streaming Video: Key Findings from the National Survey." *Against the Grain* 26 (5). https://doi.org/10.7771/2380-176X.6852.

Fleming, Rachel. 2015. "Looking at the Library Collections Budget: Budget as Story and Interrogating Narrative Frames." *Serials Review* 41 (3): 173–5. https://doi.org/10.1080/00987913.2015.1066943.

Flowers, Janet L. 2004. "Specific Tips for Negotiations with Library Materials Vendors Depending upon Acquisitions Method." *Library Collections, Acquisitions, and Technical Services* 28 (4): 433–48. https://doi.org/10.1016/j.lcats.2004.08.003.

Foley, Marianne. 2018. "Streaming Video PDA: Brace Yourself, Usage Is Coming." *Proceedings from the Charleston Library Conference*. https://doi.org/10.5703/1288284317017.

Fraser-Arnott, Melissa. 2022. "Exploring Public Library Identity Through Mission Statements." *Public Library Quarterly* 41 (3): 236–56. https://doi.org/10.1080/01616846.2021.1893568.

Frederick, Jennifer and Christine Wolff-Eisenberg. 2020. "Academic Library Strategy and Budgeting During the COVID-19 Pandemic." Ithaka S+R. https://doi.org/10.18665/sr.314507.

Garcia, Ed. Interview with the authors, September 22, 2023.

Handman, Gary. 2010. "License to Look: Evolving Models for Library Video Acquisition and Access." *Library Trends* 58 (3): 324–34. https://doi.org/10.1353/lib.0.0094.

Johnson, Peggy. 2018. *Fundamentals of Collection Development and Management*. Chicago: American Library Association.

Kanopy. 2020. "Streaming Video Trends in Public Libraries." White Paper. Kanopy. https://lib.kanopy.com/streaming-trends-in-pl-2020/

Laskowski, Mary S. 2004. "Stop the Technology, I Want to Get Off." *The Acquisitions Librarian* 16 (31–32): 217–25. https://doi.org/10.1300/J101v16n31_17.

Pan, Denise and Yem Fong. 2010. "Return on Investment for Collaborative Collection Development: A Cost-Benefit Evaluation of Consortia Purchasing." *Collaborative Librarianship* 2 (4): 183–92. https://doi.org/10.29087/2010.2.4.06.

Parker, David. 2015. "Blurring the Lines: Libraries and Video: Measuring ROI Really, How?" *Against the Grain* 27 (1): 72–3. https://doi:10.7771/2380-176X.7022.

Prelitz, Keri. 2022. "Revisiting the Streaming Video Patron-Driven Acquisition Program." *Journal of Electronic Resources Librarianship* 34 (4): 323–31. https://doi.org/10.1080/1941126X.2022.2133919.

Rodgers, Wendy. 2018. "Buy, Borrow, or Steal? Film Access for Film Studies Students." *College & Research Libraries* 79 (4): 568–91.

Rodgers, Andrea. Interview with the authors, October 10, 2023.

Spratt, Stephanie J., Gabrielle Wiersma, Rhonda Glazier, and Denise Pan. 2017. "Exploring the Evidence in Evidence-Based Acquisition." *The Serials Librarian* 72 (1–4): 183–9. https://doi.org/10.1080/0361526X.2017.1321901.

Tanasse, Gisèle. 2021. "Implementing and Managing Streaming Media Services in Academic Libraries." *Choice 360 White Paper*. Association of College and Research Libraries. https://www.choice360.org/research/implementing-and-managing-streaming-mediaservices-in-academic-libraries/.

Urban, Sandra Gall. 2022. "Evaluating Kanopy Access Models in Academic Libraries: Balancing Demand and Budget Constraints." *Collection Management* 47 (1): 37–48. https://doi.org/10.1080/01462679.2021.1940410.

Wang, Jian and Elsa Loftis. 2020. "The Library Has Infinite Streaming Content, but Are Users Infinitely Content? The Library Catalog vs. Vendor Platform Discovery." *Journal of Electronic Resources Librarianship* 32 (2): 71–86. https://doi.org/10.1080/1941126X.2020.1739818.

Wyatt, Neal. 2022. "Collection Rebalance." *Library Journal* 147 (7): 28–31.

Yale University Library. 2023. Collection Development Statement: Video Collection.

4 Licensing Streaming Video

Introduction

Whether they realize it or not, most people deal with license agreements for online services every day of their lives. Anytime a new app is downloaded to their smartphone, laptop, or other electronic device, the user is often presented with a checkbox where they affirm that they agree to the terms and conditions of the application developer. This type of agreement is often referred to as a "click-thru" agreement, and indeed, users more often than not click away without a second thought as to what is in the fine print. A 2017 survey confirmed that over 90 percent of users accept these terms of service without even reading what they just agreed to (Cakebread 2017). Frequently, unbeknownst to them, users are supplying their contact, demographic, and sometimes even geospatial tracking information to the company providing the app. In the case of popular social media platforms such as Instagram and Facebook, the companies retain the right to copy, modify, and share users' personal photos, and can transfer this right to other third parties (Thompson 2015). Not only that, the companies can unilaterally change the terms at any time and users are obligated to defend these social media platforms in court against any lawsuits while also being barred from entering any class action lawsuits against the platform.

No doubt many social media users are cognizant of at least some of what they are acceding to when they click "yes" on a platform's terms of use, even while the fine print of these terms is written to be intentionally complex to dissuade a close reading. Clearly, they are willing to accept the terms as a means to gain access to a particular app or the wide reach of a massive social network. However, when it comes to terms of service for online resources that a library licenses and pays for, it is of vital importance to give a thorough reading when considering whether the library wants to move forward with acquiring it or not.

This chapter will outline best practices for licensing electronic resources, focusing on aspects of the process that are unique to the streaming video format. While the previous chapter on budgets touched upon the negotiation of costs, this chapter will cover negotiating terms of use. Areas of note include fair use for education, the specification of public performance rights (PPR), and considerations for local hosting of video files. License agreements are foundational documents that spell out how content can be used and what uses are prohibited. Licensing will be referred back to in subsequent chapters throughout this book, as the terms of the agreement will impact access and discoverability, management of locally held files (when applicable), accessibility, and assessment of streaming video collections.

Basics of Licensing E-Resources

Licensing is one of the most critical components of the electronic resources life cycle, as it sets out the terms of use for an e-resource being acquired by a library. A license is a binding and enforceable agreement between a publisher or content owner (the licensor) and a library or institution acquiring the licensor's content (the licensee). The license agreement should clearly establish what constitutes acceptable use of the e-resource as well as any other important terms and conditions. Oftentimes, the license agreement will also address how disputes between the licensor and licensee can be resolved.

While the traditional conception of a license is a contract of substantial length, loaded with legally dense verbiage and countersigned by both parties, this does not always have to be the case. In fact, a license can be as brief as both parties would like, so long as it addresses the terms of importance to them. The language of the license need not be complex either; it is of paramount importance that both parties have a clear understanding of the license terms, so plain-speak that avoids ambiguity is preferable. In recent years, there have been efforts to make license agreements less daunting to libraries. There are numerous model licenses that exist, developed by organizations such as the Center for Research Libraries (Center for Research Libraries 2015), the Canadian Research Knowledge Network (CRKN 2022), Jisc Collections (Jisc 2018), the NERL consortium (NERL 2019), and the California Digital Library (California Digital Library 2018). These models provide boilerplate contract language designed to favor libraries with preferred terms for educational use. While not all terms within the model licenses may be applicable to certain e-resources, they provide samples of verbiage that libraries can suggest licensors incorporate into their contracts. In some cases, a contract may not be needed at all. The National Information Standards Organization (NISO), an organization that develops and maintains technical standards, created the Shared E-Resource Understanding, or SERU, as a mutually acceptable set of terms for libraries and publishers to utilize in place of a signed license agreement (NISO SERU Standing Committee 2012). To participate, a library or publisher simply needs to join the SERU registry; as long as both parties are participants, they can opt to utilize SERU in lieu of a license.

These are just some examples of tools developed to facilitate—or in some cases bypass—the negotiation process. The goal of the negotiation process, for both parties, is to arrive at license terms that are mutually acceptable. In reality, this outcome is far from assured and may be quite lengthy, depending on what terms are considered vital by either party. Reaching the conclusion of a negotiation may entail a fair amount of compromise. For librarians, it can be common to view negotiations with some amount of anxiety, as there is no assurance that a licensor will agree to all of the library's preferred terms. However, librarians should consider license negotiations as their best opportunity to advocate for the terms of use that will benefit their users. To the degree that a library carries any leverage with a licensor, it is prior to agreeing to acquire a resource. The negotiation process is the best opportunity for libraries to exert this leverage and put terms of use in place that will best serve their community of patrons. After the agreement is in place and the resource has been purchased, the library's leverage is significantly lessened. While it is always possible to modify existing licenses through a later amendment or addendum, there will be far less incentive on the part of the licensor to do so.

In terms of licensing streaming media, many of the guiding principles of negotiation apply the same as they would for any other type of e-resource. However, there are a number of additional considerations unique to the streaming format that should be weighed when negotiating an agreement. This chapter will identify those considerations and outline best practices that can be applied in negotiations. But first, some context will be provided by reviewing needs that should be addressed when licensing any electronic resource.

Red Flags and Must Haves

Overall, the majority of general licensing principles for e-resources can be directly applied to streaming video. Libraries will want to advocate for the same level of user rights that they would for non-streaming resources and will want to be aware of the same potential deal breakers when reviewing contract language. Some common "red flags" libraries may encounter in license agreements are clauses requiring indemnification of a vendor, governing law that binds the library to the jurisdiction of an outside country, or language indicating that vendors may encroach on user privacy by collecting personal data. Conversely, libraries will also want to ensure that the license covers user rights through the inclusion of language that defines all of an institution's prospective authorized users and perpetual access rights, and that the platform adheres to current guidelines for user accessibility such as Web Content Accessibility Guidelines (WCAG 2.1) (World Wide Web Consortium 2018). If such "must-haves" are not included in the agreement, the library will often pursue further negotiation to insert them into the license. Local practice can vary widely by institution and may be determined by operational status, for example, if an institution is public or private. Documenting institutional practice through assistive tools such as checklists and model license language can be highly useful throughout the licensing process.

Given that it may take a good deal of back and forth between a vendor and a library to arrive at mutually acceptable terms through the addition, modification, or removal of language from a license, negotiation is often an iterative process. Depending on the organizational structure of the library and the vendor, this dialogue may be occurring among multiple staff members across multiple levels of hierarchy. The vendors themselves may have a separate legal department that reviews all changes proposed by the library, particularly in the case of larger vendors. In order to ensure that a licensed product will meet both the user experience and technical needs of the institution, the library may have the license reviewed by information technology staff, an office of accessibility, a procurement or business office, and legal counsel, to name just a few impacted areas.

Streaming-Specific Licensing Considerations

Many of the general principles of licensing and negotiation for e-resources will apply to streaming video. However, there are also concerns unique to the format that can potentially impact the licensing process.

Communication

The vendor landscape for streaming video is highly varied, with scores of content providers, ranging from companies that license titles numbering in the tens of thousands, to smaller providers that specialize in niche areas and license select titles. Some providers may be part of a larger company that licenses content to libraries in other formats such as databases and e-books, while other providers may solely license streaming video. One particularly unique aspect of working with streaming vendors is that libraries may also deal directly with the actual filmmakers themselves when licensing independent features.

Given that streaming vendors may be organizationally smaller than other e-resource content providers, or in some cases may in fact be individual filmmakers, a license agreement need not necessarily resemble the multi-page contracts common to library e-resources. Streaming providers are often less likely to offer libraries the type of boilerplate contracts that larger library vendors frequently present up front when licensing their content. In fact, such an agreement may be a daunting proposition for an independent filmmaker not inclined to enter into a complex contract so that their work may reach its audience. When communicating with these providers, it is important to note these considerations in the process of negotiating a mutually acceptable agreement. Ensuring that all parties are comfortable with the nature and scope of the license agreement is essential. To that end, the license can be less than a single page in length, so long as it encompasses all of the usage rights that the library needs to secure. While it is true that a fully executed agreement, countersigned by both parties, is generally preferred, it is not a strict necessity to move forward with acquiring streaming content. The terms of the license do not necessarily need to have signatures; in fact, the license terms can be included within the invoice for the streaming content, so long as the language for the terms is agreed to ahead of time. The library may request a pro forma, or sample copy of the invoice in advance, to inspect and approve the language. Some form of documentation for the library to archive is important, and the library will want to avoid any click-thru terms, where the transaction happens entirely online and the library agrees to terms at the point of purchase. With that said, the license documentation can be as simple as the email thread between the library and the vendor, where the vendor affirms that they will allow the preferred terms of use that the library requests. Again, the ideal is to favor clear and plain-spoken language. In many cases, streaming video providers, particularly the smaller ones, will prefer this type of straightforward communication style and commitment to avoiding needless complexities in the licensing process.

License Term Lengths

The term, or length of a license, is another important consideration when licensing e-resources. When licensing formats such as e-journals or databases, it can be common to license annual subscriptions, and then renew the subscription after a year has passed. Sometimes a library will pursue a multi-year subscription, often to lock in favorable pricing. Perpetual access purchases are also common, typically with online archives or e-book packages.

Similarly, there can be a good deal of variation with regard to term length for streaming video. Short-term licenses for access are common, with terms of one, three, or five years being typical lengths. If multiple term lengths are available to choose from for a specific title, it will usually correspond to scaled pricing, with longer terms being associated with higher costs. Distribution rights for films often change hands over time, which is why a vendor may only offer short-term licenses in cases where they do not own the rights in perpetuity. This is frequently the case with larger streaming vendors, particularly when they offer popular feature films from major studios, such as the company Swank which licenses a shifting library of titles. Perpetual access to streaming is most likely to be available when working directly with an independent filmmaker or distributor. In contrast to studio feature films, these titles are more likely to include documentaries and short films with running times of forty minutes or less.

Access Models

Access is another important consideration that can be addressed within a license. Again, the options will typically depend on the size of the vendor. Larger vendors are more likely to offer access models where the streaming content is hosted on their platform. This is the case with major streaming vendors such as Kanopy, Swank, and Alexander Street. Larger vendors are also more likely to offer short-term licenses for finite periods of time, often limited to one year. As a result, libraries may find themselves continually renewing the license from year to year, as they would with other subscription-based e-resources. While not ideal, it may be the only licensing option available, particularly in the case of popular films produced by major studios.

Local hosting of streaming media is another potential access model. In this case, the library will receive the actual streaming media file from the vendor, and then host it on a platform where the library administers access to its authorized users. In this mode of access, the library purchases a digital site license, or DSL, for a particular title. The DSL term can be finite (e.g., one, three, or five years) or it can be perpetual, which is typically referred to as "life of file" within the contract language. The library is responsible for storing and maintaining the files themselves. The term "life of file" has some ambiguity to it, as file formats may become obsolete or there can be failures in the mechanisms for file storage. For perpetual DSL, it is a best practice for libraries to specify in the license agreement that they reserve the right to create copies of the file for backup or archival purposes, and that they can convert the file format as deemed necessary. In the case of short-term DSL, the license language may also specify that the library is responsible for deleting the file from their servers and destroying any local copies at the conclusion of the license term.

Smaller vendors may not have the infrastructure to locally host streaming files themselves on a proprietary platform. In these cases, they may opt to use third-party cloud-based solutions, such as Vimeo, for their hosting needs and provide libraries with a link to that platform instead of the actual file. In cases like this, access is often mediated through a username and password combination. Maintaining access credentials locally is less than ideal for libraries but may be the sole option with some distributors. In cases where there is username and password access, or even just a direct link with no credentials needed, it falls upon the library to ensure that only authorized users can access the streaming content. This can be done by placing the access information on a webpage behind an authentication wall, where authorized users enter their institutional credentials. Or the links can be delivered through a secure learning management system (LMS), such as ones used in electronic reserves, ensuring that access is only granted to authorized users. Commonly used LMS include Blackboard and Canvas, and are in use at educational organizations at levels, from K through 12 to colleges and universities. Maintaining access points to streaming video content will be covered in greater detail in Chapter 7. Regardless of how access is maintained, the mechanics should be clarified upfront during the licensing process, so that the library can confirm that they are able to meet the expected access thresholds. The agreement should provide clear documentation for both parties as to how access will be provided.

Preferred Clauses for Local Hosting

Given the properties inherent to the format of streaming media, some preferred terms that would be negotiated in a typical e-resource license may not be applicable. For example, scholarly sharing through a mechanism like interlibrary loan (ILL) may be appropriate for articles within an e-journal or chapter in an e-book. However, for an entire streaming film, sharing or broadcasting outside of an institution may not be viable or permissible. Most of the following preferred clauses will be applicable to cases where an institution acquires individual streaming files and hosts the content themselves. The verbiage for these suggested clauses, italicized in the below sections, comes from Yale University Library's model license for streaming video. The full model license is included in Appendix B of this book.

Converting Formats and Creating Archival Copies

> *Licensee shall have the right to transcode the content into the streaming format of the Licensee's choice and to load the digital files onto the Licensee's digital network for streaming to Authorized Users. Licensed Materials may be duplicated for the purpose of a backup copy, and Licensee is authorized to convert Licensed Materials using any medium, format, or technology now in existence or hereinafter devised, including the Internet, for the uses permitted herein.*

Ideally, when licensing individual titles where the file is locally hosted, a library will license the content in perpetuity, or for life of file. Due to the potential for failures in data storage or obsolescence of media file formats, it is not necessarily feasible that the exact digital file that is originally received by the library will continue to be the one stored and used in perpetuity. Spicer astutely notes that "it is unrealistic to expect that the exact DSL video file created or received will be the same one that will forever be stored and delivered" (Spicer 2018, 252). For continuity of access, it is a best practice to ensure the license agreement reserves the right to make backup copies of digital files for archival purposes as well as the right to convert files into other formats or re-encode them to make more manageable file sizes as necessary.

Accessibility

> *Licensor shall deliver transcripts and caption files when available. Licensee is authorized to add pop-on or closed-captioning to digital file. Nothing in this Agreement shall limit the Licensee or any end user from making lawful, non-infringing uses to facilitate access to the Licensed Materials by users who have disabilities as defined under the criteria of the Web Content Accessibility Guidelines (WCAG) 2.1. For the avoidance of doubt, the Licensor authorizes such uses.*

One key aspect of streaming media that can be addressed within a license agreement is accessibility. Accessibility is essential for providing equal opportunities to access materials for users with disabilities and is a core value of libraries. It is an area that applies both to streaming content that is licensed for use on a vendor's own platform, and for licensing streaming files to be hosted locally.

For content licensed on a vendor's streaming platform, the platform should adhere to established guidelines of accessibility. The Americans with Disabilities Act (ADA) is a common benchmark for compliance within the United States. The verbiage in the model license clause cites WCAG 2.1. WCAG is a recommended practice used internationally for websites and 2.1 is the current version as of this writing. While ADA is the enforceable legal statute within the United States, WCAG contains actionable guidelines for accessibility. The license agreement can be used to affirm the vendor's commitment to adhering to these guidelines. When full compliance cannot be guaranteed, the license should at least state that the vendor will make "best or reasonable efforts" to adhere to accessibility guidelines. Vendors should also agree to indemnify and hold libraries harmless within any disputes that may arise in a user claim stemming from a failure of the platform to be made accessible. With regard to streaming content, the platform should not only be equally navigable for all users but it should also provide closed captions for each title on the platform; Chapter 8 provides a more in-depth exploration of accessibility functionality in streaming platforms. In cases where the content on a platform is not accessible, the license should reserve the right of remediation, and allow the library to accommodate users, for example by creating their own caption files to accompany the platform content.

When licensing streaming content for local hosting, the vendor should provide textual transcripts and a caption file to accompany the streaming file, at no additional cost. Spicer frames accessibility as a moral imperative for vendors, flatly asserting: "Commercial video producers should not release uncaptioned video to the public" (Spicer 2018, 249). In cases where a vendor is unable to make accommodations for accessibility, the library should reserve the right to do so themselves and add captioning to the vendor-supplied content. The license agreement should make clear that such remediation is allowed and that it does not constitute any infringement or non-permissible modification of the licensed content.

Fair Use and Educational Screenings

> *Licensor hereby grants to Licensee the right to make the Licensed Materials available to Licensee's Authorized Users for educational, scholarly, and research purposes, and any other non-commercial use in support of Licensee's mission, as well as hold non-theatrical screenings where admission is not charged.*

For institutions in the United States, classroom screening of streaming content falls under fair use provisions of the Copyright Act of 1976 (17 U.S.C. $107)—with "face-to-face" or in-class instruction specifically being provided for in 17 U.S.C. $110(1)—though it is still a best practice to have this acknowledged in the license agreement. The agreement should explicitly state that the licensed content is being made available to authorized users for educational and non-commercial use, including teaching, research, and other scholarly purposes. Vendors may include language with restrictions going beyond US copyright law, in which case fair use could be superseded by the contract. The license should be thoroughly reviewed to identify and remove such restrictions. Additional verbiage can be inserted to ensure that nothing within the contract can be interpreted to diminish or restrict authorized usage under US copyright law.

If possible, authorized use should be negotiated to include non-theatrical screenings open to authorized users where admission is not charged. This could be applicable to campus screenings held by a student organization or a film society for example. Depending on the anticipated use of a streaming title, the library may consider negotiating for this category of authorized use, often known as PPR. Towery, Price, and Cowen note that "PPR is often sought for extra protection for the university and to cover possible community showings" (Towery, Price, and Cowen 2019, 12). In actual practice, libraries may encounter inflexibility on the vendor's part with regard to PPR. Some vendors may insist on license language that explicitly forbids public performances. Or PPR may only be included as an add-on option, often at a significant cost, and potentially with further provisions, such as being limited to a one-time campus screening. Other common provisions applied to PPR include limits on audience capacity, for example, limiting screenings to audiences of fifty attendees or less. There is a lack of an industry-defined standard of PPR, and PPR terms can vary widely among different streaming video providers. Larger providers are often more likely to restrict PPR or require additional charges. This is not always the case, however, a significant amount of content licensed through Kanopy includes PPR. When working directly with filmmakers and independent distributors, there may be more flexibility to include PPR within a license agreement. Libraries can always attempt to negotiate for PPR but should consider anticipated future use to gauge how necessary it is and if additional costs are justifiable. In the end, they may have authorized use limited to classroom instruction and individual scholarly and research purposes.

What constitutes a "fair use" for streaming video will doubtlessly continue to challenge libraries within the United States so long as the current laws persist. Spicer summarizes the situation and its impact on individual libraries thusly: "Given the purposeful ambiguity of copyright law and fair use, ultimately these decisions are guided by a certain degree of institutional, organization, and sometimes even individual risk assessment" (Spicer 2018, 254). Due to this uncertainty, libraries are frequently on their own and more prone to operating clandestinely. Cross observes that "uncertainty about the law and fear of litigation leave many librarians feeling compelled to work *sub rosa,* keeping their heads down in hopes that they won't be discovered" (Cross 2016, 2). The prevailing wisdom among libraries—at least the ones that share their practices—is erring toward a conservative approach when it comes to making streaming content available for any type of screening beyond individual or classroom use. This is wholly understandable, particularly when dealing with popular feature films and potentially litigious major studios. The emergence of increasingly online or hybrid instruction—a burgeoning model even prior to the COVID-19 pandemic—has only served to complicate the concept of the classroom and what constitutes "in-class" learning further. The library at Texas State University (TSU) developed a workflow for acquiring streaming video for use in face-to-face, hybrid, and online courses, guided by a detailed decision tree with a firm grounding in existing copyright and fair use law (Towery, Price, and Cowen 2019, 7–8). Tools like this are a welcome service for libraries unsure of how to wade into the often uncertain waters of streaming video acquisition. The librarians at TSU acknowledge their model's potential overreliance on licensing with a resistance to the application of fair use, foregoing potential budgetary savings (Towery, Price, and Cowen 2019, 14–15). While this conservative approach can be frustrating, the presence of a license agreement with concrete language on allowed uses of streaming video remains the best measure for safeguarding the interests of the library.

Digital Site License with a Physical Component

When negotiating DSL rights for individual streaming titles, the most common use case results in the vendor providing a streaming media file to the library. However, libraries may also license the right to locally digitize the file themselves. In this case, a library will typically rip and encode a streaming file from a physically held DVD. In some cases, the DVD may already have been purchased, and the library is subsequently securing the DSL permissions from the rightsholder. In other cases, the library may be purchasing both the physical DVD and the right to rip a streaming file simultaneously. Purchasing a DVD with DSL rights as a bundle will almost certainly be more costly than an individual purchase of either, with the DSL often comprising the bulk of the cost. In some instances, the acquisition may solely be for the DSL with the DVD included as a "bonus"; Chapter 5 outlines approaches for order and receipt of these bundled purchases. This type of acquisitions model is similar to how publishers would often bundle print and electronic journal subscriptions and market them to libraries. The electronic version is the cost driver, with the physical copy provided as something of an afterthought.

When purchasing physical DVDs, licensing the rights to digitize files for local hosting may be prudent for libraries seeking to build streaming collections. Libraries may also seek to negotiate PPR with the rightsholder, which is sometimes included or available as an "add-on" to purchase. Libraries should also be cognizant that smaller vendors levying additional PPR fees may be more dependent on these revenues; there is a balance to be had between negotiating for library usage rights and supporting independent distributors in the streaming marketplace. Again, when additional costs will be incurred the library should consider future use and what is justifiable within budget constraints. Such negotiations are also more likely than not only available for smaller, independent features when working directly with the filmmaker or distributor. Licensing rights for features from major studios are almost certain to be exponentially complex and cost-prohibitive, if even possible at all.

Netflix and Other Subscriber Platforms

One newer area of licensing complexity that has arisen in recent years is the rapid growth of titles exclusive to subscription platforms such as Netflix, Hulu, and Amazon Prime. These platforms have thrived off of individual user subscription models and show no indication of moving toward making sitewide educational licenses available to libraries. Netflix has made some limited exceptions in the case of certain documentary content (e.g., *The Social Dilemma*, *Crip Camp*), which include a grant of permission for educational screenings. These rights allow a Netflix account holder—in most likely cases a course instructor—to present a one-time screening of a Netflix documentary in a classroom or educational setting. The user must verify that the title includes permission for educational screening, which can be found on the film's details page within the Netflix Media Center on the platform (Lambert 2022). The educational grant includes a number of stipulations on the use of the title in a classroom setting, with clarification of "one-time use" applying to once per semester if an instructor seeks to include a title in a recurrent course. It is also highly unlikely that an entire film can be viewed in one class session thus making it even more difficult to use Netflix content in courses. The allowable titles comprise a very limited subset of the documentary content on Netflix and do not include non-documentary series or films that may potentially be used in a course (e.g., *Stranger Things*, *Roma*). Furthermore, the vast majority of Netflix content is not available on DVD/Blu-ray, leaving the library with no options for acquiring it in any format. As a result, instructors who may want to incorporate content from Netflix—or other commercial platforms—as course materials often have to tell their students to subscribe to the platforms themselves. Such solutions are less than ideal; they shift the burden upon students to shoulder additional costs to potentially multiple streaming platforms per semester, at a time when the affordability of textbooks and other educational materials is already alarmingly high. Furthermore, as some librarians have noted, it does not align with the mission of libraries to build permanent collections (Handman 2010; King 2014).

A good amount of streaming content ends up on freely available platforms like YouTube. In many cases, these titles are not posted by the actual rightsholders. Content not posted by rightsholders is always subject to takedown without notice and may have been intentionally pirated. Libraries should make every effort to source and license streaming content from the rightful owners of the content. Some video content may truly be an "orphan" work with no known rightsholder, but libraries should first exhaust all possibilities to license the content prior to that determination.

When working with instructors who want to include "free" videos of orphan works, librarians should advise them that the title may not continue to remain there.

Conclusion

The challenges of licensing streaming media are quite similar to those of all e-resources, and license agreements for streaming are likely to share many common clauses. The differences specific to streaming formats most often factor into the access method, particularly when digital files are being acquired for local hosting. Terms for streaming media tend to be finite, with single-year licenses similar to other subscription-based e-resources. Perpetual access, or life of file licenses, can sometimes be negotiated directly with filmmakers and distributors but are often quite costly on an individual title basis.

As with any other e-resource license, the same principles of negotiation should be applied by libraries to streaming agreements. The contract language should clearly spell out what falls under authorized use for the content. The library should remove any clauses which they find to be unenforceable. Overall, the anticipated use of the content should be in mind when negotiating the license. Areas of interest such as PPR may be brought into the negotiation, with a recognition that such grants of permission may ultimately only be available at increased cost to the library. The library should be guided by how their streaming collection can best meet the needs of its users, and then negotiate accordingly.

Streaming Vignette—Lyrasis

Name of the library: Lyrasis

Reporting librarian and title: Michael Rodriguez, Senior Strategist, Content and Scholarly Communication

Type of library: Nationwide consortium

User community: Works on behalf of over 1,100 institutional members comprising libraries (including university, community college, and public libraries), archives, and museums

Total streaming budget: Not applicable

Organizationally, Lyrasis licenses content on behalf of its 1,100-plus members, negotiating group license agreements that any member can choose to sign onto. This is a beneficial service to all members, particularly smaller libraries that may not have the staffing or bandwidth to individually negotiate and license their own content. As a consortium representing a large community, Lyrasis is able to leverage the size of its membership base to negotiate with content providers and achieve favorable pricing and terms and conditions of use.

In the realm of streaming video, Lyrasis has negotiated deals with many of the larger content providers, including Kanopy, Alexander Street, Ambrose Video,

Bloomsbury, Docuseek, and Swank. In seeking providers to partner with, Lyrasis takes scalability into consideration and focuses on which partnerships will deliver the greatest overall benefit for the most members. Rodriguez notes that streaming collections can often be "boutique" at many libraries, which is to say, many acquisitions are on an individual title basis and done through smaller providers. Lyrasis instead focuses on negotiating deals for full collections from larger providers, to yield greater benefits and uptake from members.

Along those lines, in 2022 Lyrasis negotiated a deal for Kanopy BASE (Bundled Academic Subscription for Education), Kanopy's bundled package of 10,000 titles for educational use. Distinct from Kanopy's patron-driven acquisitions (PDA) plans, Kanopy BASE offers predictable pricing for libraries with a one-time annual subscription fee. A 10 percent discount was negotiated for all participating members, which gets applied to a tiered pricing structure based on institution size. Members who have already negotiated licenses with Kanopy can still prioritize their existing terms while taking advantage of the Lyrasis discount.

Much of Lyrasis's strategy in working with content providers is based on a set of core principles that values authenticity, integrity, inclusivity, connectivity, and sustainability. To support these principles, Lyrasis looks toward fostering partnerships with providers that are creative or innovative. For Kanopy BASE, the negotiations occurred in the early planning stages of the model, allowing Lyrasis to exert influence on the model's development and leverage member feedback on what would be most beneficial to libraries.

In terms of the licensing process at Lyrasis, a team-based approach is utilized, with duties spread across four senior strategists and several additional staff who divide up responsibility for a portfolio of over 100 vendors. Strategists utilize an internal checklist during the licensing process and can also share questions with each other through a Slack channel. Once mutually acceptable terms are arrived at with a content provider, the Lyrasis signatory will give the agreement a final review before the contract is fully executed. For contract renewals, Lyrasis generally seeks to impose a cap of 3 percent or lower on any pricing increases. When proposed increases come in higher, an evidence-based approach is used to push back on providers, pulling usage statistics, enrollment trends, and other data from subscribing members as needed. License renegotiations occur on a periodic basis, with a goal of refreshing each central product license every five years to ensure that terms do not go out of date and remain responsive to member expectations.

When asked to identify the greatest challenges of the streaming video format, Rodriguez pointed to the lack of scalability in pricing models, explaining that reasonable pricing for single-title acquisitions is still an elusive area. Noting parallels to the public library e-book market, he also cites the move toward exclusively access-based acquisition models as a problem. The stability of content is another challenge, as titles are regularly pulled from platforms and re-packaged in other collections or by other providers, in much the same way that is done on commercial streaming platforms. The issue of preservation comes up as well in life of file perpetual purchases, as format standards for video can change regularly. As a format, streaming video has proven resistant to at-scale solutions for shared preservation and interlibrary lending of content. Open-access programs and initiatives—many of which are community-based, with a large number supported by Lyrasis—have yet to materialize in a notable way for streaming video. More broadly speaking to the role of consortia, Rodriguez notes that licensing and negotiation are decreasingly a priority area, as consortia move away from this service model into ones more focused on maintaining shared systems and infrastructure as well as other specialized and strategic areas of work. As a result, libraries may continue finding they have to go it alone when building streaming video collections.

Lyrasis Program Overview, Core Principles: https://www.lyrasis.org/content/Pages/program-overview.aspx

References

Cakebread, Caroline. 2017. "You're Not Alone, No One Reads Terms of Service Agreements." *Business Insider*, November 15. https://www.businessinsider.com/deloitte-study-91-percent-agree-terms-of-service-without-reading-2017-11?op=1.

California Digital Library. 2018. CDL Model License Agreement. https://cdlib.org/wp-content/uploads/2019/02/CDL_Model_License_2018.05.14_public.docx.

Canadian Research Knowledge Network. 2022. CRKN Model License. https://www.crkn-rcdr.ca/en/model-license.

Center for Research Libraries. 2015. LIBLICENSE Model License Agreement. http://liblicense.crl.edu/wp-content/uploads/2015/05/modellicensenew2014revmay2015.pdf.

Cross, William. 2016. "More than a House of Cards: Developing a Firm Foundation for Streaming Media and Consumer-Licensed Content in the Library." *The Journal of Copyright in Education and Librarianship* 1 (1). https://doi.org/10.17161/jcel.v1i1.5919.

Handman, Gary. 2010. "License to Look: Evolving Models for Library Video Acquisition and Access." *Library Trends* 58 (3): 324–34. https://doi.org/10.1353/lib.0.0094.

Jisc Collections. 2018. Jisc Model Licenses. https://subscriptionsmanager.jisc.ac.uk/about/jisc-model-licence.

King, Rachel. 2014. "House of Cards: The Academic Library Media Center in the Era of Streaming Video." *The Serials Librarian* 67 (3): 289–306. https://doi.org/10.1080/0361526X.2014.948699.

Lambert, Isadora. 2022. "Netflix Documentaries that are Available for Educational Screenings." *Video Librarian*, February 23. https://videolibrarian.com/articles/lists/netflix-documentaries-that-are-available-for-educational-scr/.

NERL. 2019. NorthEast Research Libraries Consortium General License Agreement for Electronic Resources. https://nerl.org/wp-content/uploads/2021/03/NERLModelLicense-_61019_a.pdf.

NISO SERU Standing Committee. 2012. SERU: A Shared Electronic Resource Understanding. https://groups.niso.org/apps/group_public/download.php/8593/RP-7-2012_SERU.pdf.

Spicer, Scott. 2018. "Perspectives on the Role of Instructional Video in Higher Education." In R. Hobbs (ed.), *The Routledge Companion to Media Education, Copyright, and Fair Use* (pp. 236–57). New York: Routledge.

Thompson, Cadie. 2015. "What You Really Sign Up for When You Use Social Media." CNBC.com, May 20. https://www.cnbc.com/2015/05/20/what-you-really-sign-up-for-when-you-use-social-media.html.

Towery, Stephanie S., Amanda N. Price, and Karen E. Cowen. 2019. "Video Streaming Licenses: Using a Decision Tree and Workflow Chart." *The Journal of Copyright in Education & Librarianship* 3 (1). https://doi.org/10.17161/jcel.v3i1.7483.

World Wide Web Consortium. 2018. Web Content Accessibility Guidelines (WCAG) 2.1. https://www.w3.org/TR/WCAG21/.

5 Acquiring Streaming Video

Introduction

Once a license agreement has been executed, the formal acquisition process for streaming video content can take place. While this process can be—and typically is—straightforward, there are a number of factors unique to the streaming format that can impact how the ordering and receipt of content occurs. Receiving streaming video content can be as simple as the content provider switching on access once the order and payment are received. However, locally hosted files will entail some sort of file transfer or download process. Additionally, digital site licenses (DSL) for locally hosted streaming media may be bundled in some way with physical media formats like DVD/Blu-ray. This chapter will unpack the potential variations in workflow for ordering and receiving streaming video.

Vendor-Hosted Streaming Content

In terms of ordering and receipt, the most straightforward streaming acquisitions are for content that is hosted by the vendor or publisher on their own platform. These acquisitions will be entirely, or almost entirely, similar to orders for other vendor-hosted electronic content, including databases, e-journals, and e-books. The same fundamental principles of receipt of content will apply to streaming, with the vendor supplying access to the content at the point of payment.

Like other e-resource formats, content providers may employ a number of different acquisitions models for streaming content, with subscription being among the most common. A library will subscribe to a streaming title—or more likely, a collection of titles—for a defined period of time, most frequently one year. Effectively, the library is leasing the content, and will not retain ownership once the subscription term is over. Post-cancellation access to the content that was available during the subscription term, which vendors often provide for e-journal content, is not a consideration for streaming video content. Once the subscription period is over, the library will need to renew the subscription or access to the content will be lost.

Another frequently employed acquisition model for streaming video is patron-driven acquisition (PDA), sometimes also referred to as demand-driven acquisition (DDA), where purchases are triggered at the point of need. Mediation adds an extra layer of security for library budgets to protect against a PDA plan quickly incurring costs. In the second half of the last decade, as licensed streaming video content exponentially grew in availability, utilization in education, and popularity among users, libraries quickly found that PDA plans were depleting budget allocations at alarming rates (Streeter, Kennedy, and Saragossi 2023; Lowe et al. 2020). As a result, librarians who manage streaming media budgets have expressed the need for thorough and consistent monitoring of PDA budgets, as well as the desire for vendors to offer more granular controls on purchase triggers (Roll 2019).

Evidence-based acquisition (EBA) is another model provided by some streaming vendors, where an upfront amount is committed by the library to secure access to a pre-selected package or collection of streaming titles. A noteworthy difference from PDA is that for EBA, typically the purchases are for perpetual access licenses on the vendor's platform; Chapter 2 contains a more in-depth analysis of these acquisition models. Both PDA and EBA offer distinct benefits and necessitate different workflow approaches for libraries, which will be covered in the next section of this chapter.

Ownership versus short-term licensing is an important consideration when ordering streaming video. It will impact how workflows are managed both on a one-time and ongoing basis. Similar to subscriptions, PDA purchases are typically only available for limited periods, usually for one year after the trigger date. At that point, access will expire and the title will only be available to the library again through another user's trigger request. Short-term licensing such as this, whether done through PDA plans or otherwise, results in a need for more ongoing order maintenance by library staff, in much the same way that a database or e-journal subscription would. In some cases, individual titles or collections of streaming titles may be available for perpetual purchase on a vendor-hosted platform. However, subscription and PDA are the more dominant models for vendor-hosted content, with perpetual access being far more common with locally hosted DSL titles that are acquired for the life of the file.

Ordering and Receiving Vendor-Hosted Streaming Video

Ordering streaming video that is being hosted by the vendor on their own platform is a fairly straightforward process that mirrors the ordering workflow for other e-resource content such as databases, e-journals, and e-books. Let us start by looking at subscription-based or short-term licensed streaming video content. Picking up from the completion of license negotiation and the execution of the license agreement, as detailed in the previous chapter, the overall ordering workflow steps can be outlined as such:

- Purchase order (PO) is created in the library system and formal communication of the order to the vendor;
- Invoicing and payment process;

- Access is switched on and notification of availability is provided;
- The conclusion of the license term prompts a renewal decision.

As seen in Figure 5.1, the order workflow is fairly linear, with the vendor providing discovery records and access at the point of payment receipt. The overall workflow is more or less the same whether it is a single streaming title being ordered or a collection of titles. As with other subscription-based electronic resources, there is the potential for a cyclical renewal process at the expiration of the initial license term. Unlike subscribed journals, however, substantial amounts of new content are not being added year-after-year. The library is essentially renewing a lease on the content. Packages may see individual titles being added, or dropped, but a subscribed title or collection is going to remain largely the same each year, which makes it important to be aware of what the institutional need is. Will a specific title be used and re-used in an annually taught course, or is it a specialized request? This prior knowledge can determine whether an order is set up like a continuing resource or a one-time purchase.

Patron-Driven Acquisition

Workflows for PDA plans are subject to substantially higher degrees of variability, depending on local practices for order maintenance, establishing payment and moderation preferences with vendors, and the mechanics of how access is triggered. A general workflow for PDA orders would be as follows:

- Library formalizes agreement terms with vendor: initial budget or deposit amount (if required by vendor), trigger mechanism, length of license terms;
- PO is created in the library system according to local practice: continuing order, deposit account, or series of one-time orders;
- Access to individual titles provided at trigger point;
- Invoices or account statements are sent on a periodic basis (e.g., monthly, quarterly).

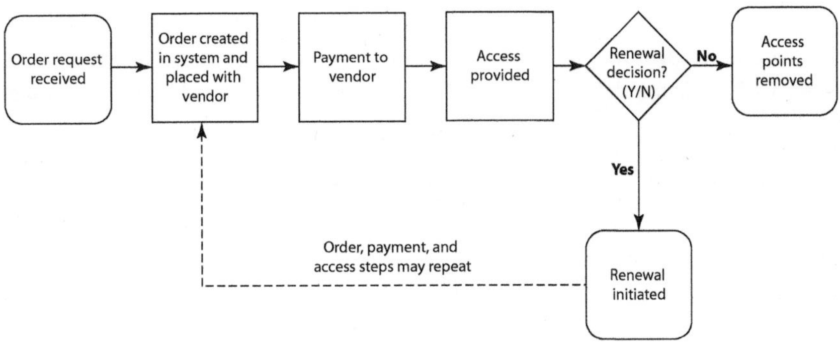

Figure 5.1 Order and receipt of vendor-hosted streaming video.

Notably, PDA plans can entail extensive negotiations with vendors to formalize terms before any order setup occurs. How the order is set up in a library's system will be influenced by local practice. It could be set up as a continuing order or a deposit account; alternatively, each new contribution to the PDA fund could be treated as a separate order. Once the plan is active, access is on demand and instantaneous. Depending on the vendor and preference of the library, payment may be made upfront or as titles are activated. Many PDA vendors will require at least an initial deposit of funds at the onset of the plan. Figure 5.2 maps the process of ordering, payment, and access for PDA models that utilize upfront deposits.

Invoices or billing statements are provided by the vendor on a periodic basis (e.g., monthly, quarterly) to allow for internal accounting. Figure 5.3 maps an open or "pay-as-you-go" variation of PDA where invoices are generated after access is triggered.

The variability of PDA also applies to license terms and the renewal process. For individual titles, the license term begins (typically for one year) at the date when the purchase is triggered. As a result, the availability of access to titles is on a rolling basis, with new titles constantly being triggered and license terms for previously available titles expiring. Continual maintenance of access points poses a unique set of challenges, which will be addressed in greater detail in Chapter 7. The asynchronous start dates for PDA titles are in contrast to the subscription-based acquisition of a streaming collection, wherein all titles within the package fall under the same start and end date of the subscription term. As demonstrated in Figure 5.2, the potential endpoint for a PDA plan funded by upfront deposits is when the funds expire, resulting in a decision point to continue the plan with more funds (although there could be a "top up" of funds at any point prior to expiration). With the "pay-as-you-go" variant in Figure 5.3, the endpoint is more nebulous, as invoices are being issued periodically and direct action is required from the library to notify the vendor of their intent to conclude the plan.

As an additional note of relevance to public libraries, some content providers offer variations on patron-driven models, known as pay-per-use (PPU) or cost-per-circ. While the basic mechanics are the same for accessing content on the user end, the costs of the plan are assessed in individual uses, views, or circulations ("circs") of a title. This is effectively a token-based model; instead of the one-year license term triggered by PDA, the library has purchased a set allotment of tokens or video plays that get spent down as users watch different videos. The overall workflow remains the same as well: the library is invoiced for an initial deposit or outlay amount upfront,

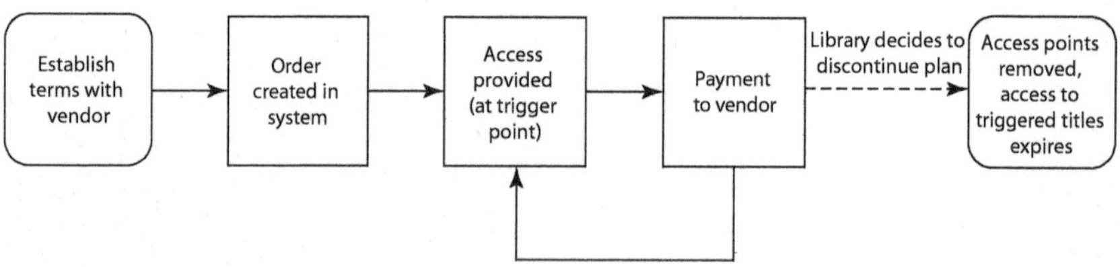

Figure 5.2 Order and receipt of PDA plan with deposit funds.

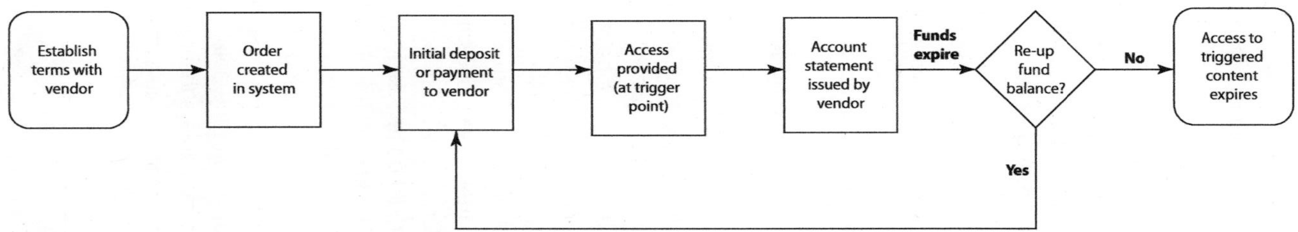

Figure 5.3 Order and receipt of open PDA plan

re-upping funds when necessary as the deposit gets spent down through usage. The library can further modify these plans in additional ways, for example by capping the amount of videos an individual user can play in a month or suppressing titles that count for multiple plays. This acquisition model is more frequently made available to public libraries, and in some cases, community college libraries.

Evidence-Based Acquisition

In EBA, a notable distinction from the workflow steps in PDA plans is that access to all of the collection titles is provided upfront following payment; no triggering of acquisition from the user end occurs. A generalized EBA workflow can be outlined thusly:

- Library formalizes agreement terms with the vendor: scope of collections and the length of plan;
- PO is created in the library system, treated as a one-time order;
- Invoicing and payment process;
- Access is switched on and notification is provided.

Workflows for EBA require an upfront agreement between the library and the publisher on the scope of the collection being made available: how many titles are to be included, subject/content areas of the title list, as well as the term of the plan, meaning the period of time that the content is being made available for. Pricing is agreed upon with the understanding that at the term's conclusion, the library's cost commitment will go to the purchase of select titles within the EBA collection. At that point, a PO can be created in the library system, and an invoice generated by the publisher. EBA would generally be treated as a one-time order, though a library would have the option to resume the plan at the term's conclusion. In doing so, it is likely that the scope of the collection would be revisited, as it would no longer include the purchased titles from the previous iteration of the plan. Once payment is received, the publisher would turn on the access to the collection and provide a MARC record set. The library would ingest the records into its catalog and set up access points through its discovery tools. As seen in Figure 5.4, the initial process steps are somewhat similar to a deposit-based PDA plan, with the notable distinction that access to *all* content for an EBA plan is made available subsequent to payment. Following the conclusion of the EBA plan's term, the library also has the opportunity to select titles

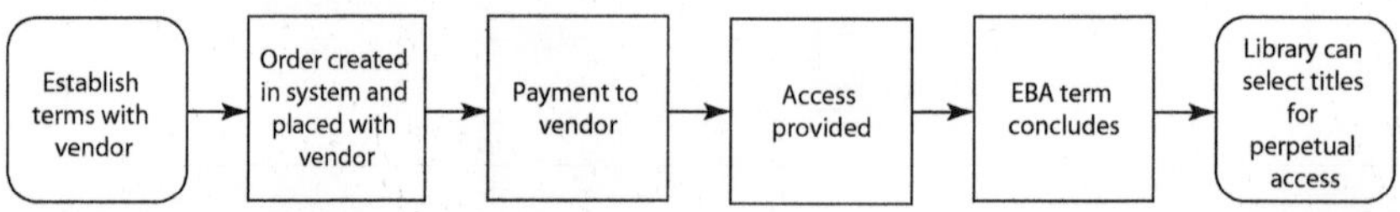

Figure 5.4 Order and receipt of EBA plan.

to acquire perpetually. Should the library opt to resume the EBA plan, the ordering process would begin anew in the same linear fashion.

Locally Hosted Streaming Content

One area of additional complexity in streaming video acquisitions is the challenge associated with titles that are purchased from a vendor or content provider with the understanding that the library itself will host the film. This is not entirely unique to streaming video resources, as there are other e-resource formats that a library may acquire and locally host. Large-scale datasets, for example, are increasingly being purchased by libraries for the purpose of data mining and analysis. While many are vendor-hosted, in some cases a library may purchase or lease the datasets to be mounted on their locally maintained platform or servers. E-books are, more often than not, hosted on a publisher platform, but in some cases, a library will license an e-book and receive a PDF or EPUB file from the publisher for local hosting. Most commonly this occurs in the case of smaller publishers, who may not have the infrastructure to maintain their own platform but can provide source files to libraries who are set up to host locally held content.

The use case for locally hosted streaming video is quite similar and often involves smaller content providers supplying video files directly to libraries. Given that there is a wide number of independent distributors and filmmakers who are frequently creating documentaries or other educational content that would support an academic curriculum, it is not surprising that these acquisitions frequently occur with libraries and are more common to streaming video than other e-resources formats. The basic principles for ordering and invoicing vendor-hosted streaming video would still apply for locally hosted videos. Providing access is where the workflow presents added complexities, as file receipt can occur in a number of ways, and cataloging and discovery is subject to local practice.

The basic acquisition workflow for locally hosted streaming video is as follows:

- Library negotiates agreement with distributor or rightsholder;
- PO is created in the library system as a one-time order (can also be a continuing order if it is known in advance that it will be renewed);
- Invoicing and payment process;
- Receipt of file;
- Ingest to local streaming server/platform;
- Creation of access points.

As was discussed in the previous chapter on licensing, one unique aspect of streaming video acquisitions is that the library often works directly with the filmmaker or rightsholder to obtain permission for local hosting. Depending on the license terms, a library may acquire perpetual rights for the life of the streaming file, or a short-term license that may be single or multi-year. In terms of the PO, that means that POs will typically be one-time orders, but may potentially be treated as continuing orders. A potential use case for the latter is a video with only a short-term license option available where the library knows that there will be a year-after-year need for access. Next, an invoice would be issued and paid, as would typically occur with vendor-hosted content. A noteworthy difference is that payment is often going directly to an individual, who may prefer to receive payment via a money transfer app such as PayPal, Venmo, or Zelle. Library business offices often are set up to make payments through checks, so if they have the flexibility to use transfer apps, it can often expedite the payment portion of the workflow. Vendor agents are another potential option when dealing with large numbers of independent distributors, as they offer libraries a single source of payment triage. While not as common as serial subscription agents, there are third-party agents that can handle streaming video, typically with service fees assessed on an individual title basis.

Once the payment is received, the library can receive the actual streaming file in a number of ways:

- File-sharing apps (Dropbox, WeTransfer, Google Drive)
- Delivery on physical storage device (USB drive, external hard drive)
- File to be ripped from a delivered or already owned DVD

Delivery via file-sharing app is the most common method of transfer, mainly for reasons of speed and simplicity. The library downloads the streaming video from a vendor-provided link (or is granted permission to download from a shared storage space). The file is then saved locally and uploaded to the library's streaming platform. In some cases, the file might be mailed on a physical storage device such as a USB drive. This is a rarer delivery method, lacking the convenience or speed of file transfer. The use case for delivery on a physical device would be for excessively large files or possibly consolidating a large number of ordered files into a single method of receipt. Following the receipt of the files, the concluding steps of the local hosting workflow are ingesting the file to the library streaming platform and the creation of access points. As seen in Figure 5.5, potential decision points arise for renewal and file retention based on the term of the license. Chapter 6 will cover the management of locally hosted files in greater detail while Chapter 7 will delve further into access maintenance.

Another potential method of receipt is where the library rips a streaming video file from a DVD that is physically delivered or already owned. The permission to do so is established when the library is negotiating a license agreement with the film distributor. A common use case is that the distributor sells the DVD to the library with a DSL included for an additional fee. The license will specify the grant of the rights to rip a streaming file for local hosting from the DVD. In other cases, the DVD had

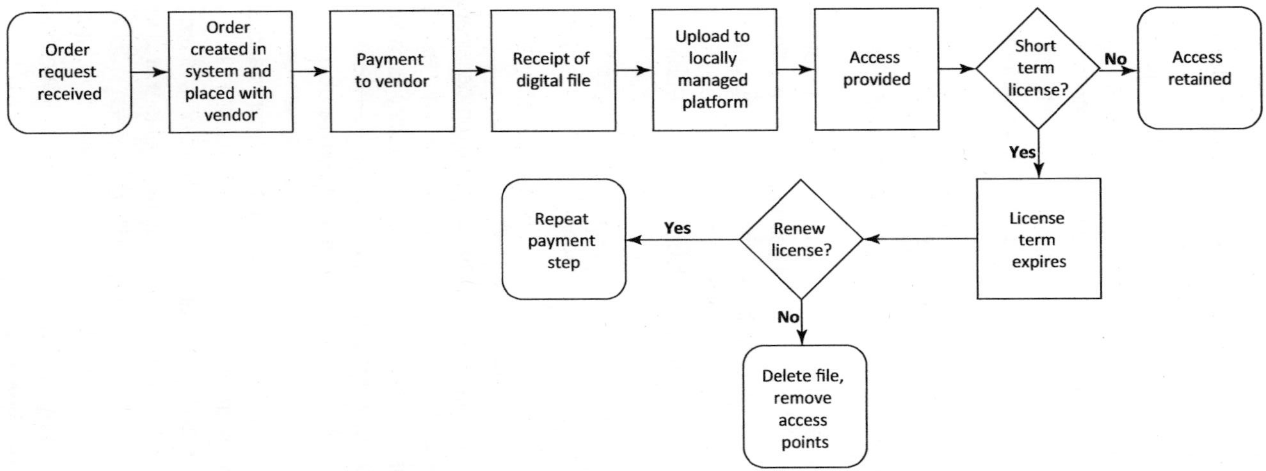

Figure 5.5 Order and receipt of locally managed streaming video.

previously been purchased and is currently owned by the library, and the DSL has been negotiated for the right to rip a locally hosted streaming file from the existing DVD.

Clearly, the preference will be for distributors to provide streaming files themselves. Ripping from DVDs requires both hardware and staff bandwidth on the part of the library. DVD optical drives could be considered "specialized" hardware, as they are not included by default on modern desktop computers and laptops. As a result, acquisitions units may have to collaborate with outside areas such as the media center or a digital preservation department in order to obtain the streaming files from DVDs. The overall question of "who does what?" when streaming video acquisitions entail both a physical and electronic component can manifest in numerous ways, depending on the use case. The following section will address these cases and present potential solutions.

Combined Streaming and Physical Video Orders

Aside from the major streaming platform vendors, many, if not most, distributors of streaming video are still in the business of selling physical media along with DSLs for streaming. Oftentimes an individual title will have multiple purchase choices available; that may include separate options for standalone DVD or streaming DSL, a DVD and DSL bundle option, and sometimes further options if Public Performance Rights (PPR) are available as a separate add-on. For library technical services units, this can quickly lead to complications, as physical materials and e-resource materials follow different workflows for payment, receipt, and access. More often than not, particularly in larger libraries, the staff who handle physical materials are in entirely different teams or units from e-resources staff.

Given the fact that staffing duties in libraries are often designated by format, it is unsurprising that workflows entailing multiple formats can lead to confusion over responsibility. Such was the case when Columbia University Libraries began assessing policies to process DVD + streaming requests for course reserves in 2012; there was initial confusion as to whether ordering and payment would be handled within the monographs or e-resources departments. Ultimately this confusion was overcome by modifying order forms to include multiple recipients across departments and documentation in a shared wiki (Koennecke, Marcin, and Pavlick 2014, 399–401). In a similar fashion, the heads of the e-resources and monographic acquisitions units at the authors' institution collaborated on shared documentation to address multi-format video acquisitions. An adapted version of that document is presented here, as seen in Table 5.1.

Table 5.1 outlines a number of different use cases for video acquisitions that include a combination of physical and online materials. It delineates between physical materials staff and e-resources staff, assigning responsibility based upon which format workflow takes precedence and how the order request originates. In all cases where online access is being provided to users, it is a best practice to have a license or agreed-upon terms of use in place, meaning that licensing staff will become

Table 5.1 Order and Receipt of Vendor-Hosted Streaming Video

Physical item only (DVD or Blu-ray disc (BD))	*License needed:* No *Who places the order?* Physical material staff *Who receives the content?* Physical material staff *Who receives the invoice?* Physical material staff *Who catalogs the content?* Physical material staff
Streaming only	*License needed:* Yes *Who places the order?* E-resources staff *Who receives the content?* E-resources staff *Who receives the invoice?* E-resources staff *Who catalogs the content?* E-resources staff
DVD/BD with streaming file	*License needed:* Yes *Who places the order?* Physical materials staff, following notification that license has been executed *Who receives the content?* DVD/BD: physical materials staff Streaming file: E-resources staff *Who receives the invoice?* Physical materials staff *Who catalogs the content?* DVD/BD: physical materials staff Streaming file: E-resources staff *Additional considerations?* If possible, purchase order and invoice should break out physical and streaming copy
Streaming file with DVD/BD included	*License needed:* Yes *Who places the order?* E-resources staff *Who receives the content?* DVD/BD: physical materials staff Streaming file: E-resources staff *Who receives the invoice?* E-resources staff *Who catalogs the content?* DVD/BD: physical materials staff Streaming file: E-resources staff *Additional considerations?* If the DVD/BD is included as a "bonus," the original selector will determine if it should be added to the collection
DVD/BD with option to rip a streaming file	*License needed:* Yes *Who places the order?* Physical materials staff, following notification that license has been executed *Who receives the content?* Physical materials staff *Who receives the invoice?* Physical materials staff *Who catalogs the content?* DVD/BD: physical materials staff Streaming file: E-resources staff *Additional considerations?* After the DVD/BD is received and processed by physical materials staff, it can be routed to E-resources staff to rip the streaming file. If possible, purchase order and invoice should break out physical and streaming copy as separate line-item expenditure types

(Continued)

Table 5.1 (Continued)

DVD/BD is already owned, option to rip a streaming file	*License needed:* Yes *Who places the order?* E-resources staff *Who receives the content?* E-resources staff; pre-owned DVD/BD is routed from collection *Who receives the invoice?* E-resources staff *Who catalogs the content?* E-resources staff *Additional considerations?* Following licensing and payment, pre-owned DVD/BD is routed to E-resources staff to rip the streaming file
Streaming file to be burned to a DVD/BD	*License needed:* Yes *Who places the order?* E-resources staff *Who receives the content?* E-resources staff *Who receives the invoice?* E-resources staff *Who catalogs the content?* DVD/BD: physical materials staff Streaming file: E-resources staff *Additional considerations?* Once the DVD/BD is burned, it can be routed to physical materials staff to process into collection

involved in some capacity for any orders that include a streaming component. While local practice will vary across libraries, with staff responsible for both physical and electronic formats at some institutions, the general assignations in Table 5.1 can be applied regardless of institutional size or organizational structure. While Blu-ray discs are sometimes included as part of physical + streaming bundles, DVD is still the more common physical format, and will be referred to throughout.

Physical or Streaming Only

These are, naturally, the most straightforward workflows for acquisition when solely one format type is being ordered. As seen in Table 5.1, these acquisitions establish a baseline and clear delineation between workflows within physical materials units and e-resource units. While streaming workflows can vary depending on acquisition model, as outlined earlier in this chapter, they will likely fall under the purview of the same e-resources staff regardless.

DVD + Streaming Combination Orders

The next two options in Table 5.1 present the same bundle of physical and electronic formats in a single order. They are separated based on how the order is initiated. In one use case, the original request is specifically for a DVD, with a streaming file included or available for an additional fee. In the second case, the requestor has asked for a streaming video, and the DVD is being included, possibly as a "bonus" at no extra charge. This type of bundling was once common for journals that had both print and electronic components. It is sometimes still practiced by publishers who only offer e-journal access with combined print subscriptions. A library may only want online access and then need to decide if the print copies will be processed or instead routed for recycling or donation. Similarly, when a "bonus" DVD is included with a streaming order, the original selector should specify if it will be retained for the collection.

Being aware of how the order request originates is useful in designating whether the physical materials unit or e-resource unit takes the lead with regard to ordering, payment, and access. While some of these workflow steps can be handled by either unit—and may in fact require the involvement of both—having a policy in place that assigns responsibility based on the originating format of the purchase will prevent initial delays over the confusion of "who does what?"

Regardless of whether the initial request is for a streaming file or a DVD with streaming included, a license or agreed-upon terms should first be negotiated with the distributor prior to ordering. In all the use cases in Table 5.1, aside from physical-only purchases, a handoff to licensing staff will occur following the initial request.

Once licensing is complete, the order will be placed by acquisitions staff within the unit responsible for the original desired format, with that unit handling the payment as well. Since licensing is not typically part of physical format ordering, an additional notification may be required to inform the physical materials staff that a license has been executed and the order can go forward. Depending on local practice, it may also be necessary to itemize the physical and streaming components of the purchase separately on the invoice. This would be the case if the funds within the library's ledger were delineated by format, with separate fund lines for physical versus electronic materials. Depending on the vendor, this may not be possible, as they may invoice all components of the physical + streaming bundle as a single line item. The designated unit responsible for ordering and payment should establish this with the content provider upfront, prior to creating the order in the library system, so the processing of payment can move seamlessly once the order is placed.

Receipt of combined physical + streaming orders will be bifurcated across the responsible format units in the library, regardless of how the order originates. Here, it is useful for the staff formally placing the order to specify both the physical shipping address for the DVD and the contact information of the e-resources staff who will be downloading and managing the streaming file component of the order. This will allow the workflows of each unit to function more seamlessly and prevent delays; for example, if the physical materials unit staff has to determine who in the e-resources unit a streaming download link needs to be forwarded to. An example email template to use with the distributor would be as follows:

> Dear **[video distributor contact]**,
> The library at XYZ University is interested in ordering **[title of resource]** on DVD with a digital site license for the streaming video file.

Please ship the DVD to the following address:
[address for physical receipt]

The download information for the streaming file should be sent to the following contact:
[email address(es) of staff managing locally hosted files]

On the invoice, please itemize the DVD and streaming DSL as separate line items. **[Note: when applicable]**. The invoice should be sent to the following contact:
[email address of staff processing invoices]

Best
[your name]

It is often a good practice to include all of the involved staff members in the email message, so they can know to expect a deliverable from the order. Once the physical and streaming material components have been received by the appropriate unit, those respective areas can carry out the workflows for cataloging and access.

DVD with the Option to Rip a File for Streaming

In this use case, there is a multi-format order with the electronic component being contingent upon additional action needed after receipt of the physical material, namely the ripping of a streaming file from a DVD. This is a case where the physical materials unit can take the lead responsibility for the acquisition, as it will be originating as a DVD purchase. Again, prior to the order being placed, there will need to be a handoff to licensing staff to first secure the right to rip a streaming file for local hosting from the distributor. Once an agreement is in place, the physical materials unit can proceed with ordering, payment, and receipt of the DVD.

At the point of receipt, it is perhaps advisable that a handoff occurs to the staff responsible for ripping DVD files, whether that happens within the e-resource unit or a separate unit equipped with the hardware to do so. Once the DVD is in the possession of technical services staff, but prior to cataloging and adding the item to the library collection, is an optimal time for ripping the streaming file. Otherwise, additional steps may need to be taken to charge out the DVD and route it to the appropriate staff, which could delay the item being made available to patrons. Conversely, the ability to track the DVD while it is in the process of having the streaming file ripped may be preferable, so the library may process it upon receipt, and suppress the item record so patrons do not see its availability yet. The ultimate decision may be dependent on local practice, library size, as well as geographic location, and physical distance of the units involved in the various handoffs for ripping a streaming file from a physical DVD. Once the streaming file is ripped from the DVD, the physical and e-resources material units can resume their respective workflows for processing and access.

A slight variation to this workflow may occur when the library chooses to acquire a streaming license for a DVD that is already owned. In this use case, the library has previously purchased a DVD and subsequently purchases a DSL from the content provider. The main difference in workflow is that the physical material has already been received and is part of the library's collection. After the license agreement has been executed and payment is made, the DVD from the collection can be routed for the streaming file to be ripped.

Streaming File with the Option to Burn to DVD

Essentially the inverse of the previous use case, this order will originate within the e-resources unit but entail many of the same considerations for handoffs among units. While this is a rarer use case, it is not uncommon for libraries to want to retain a physical preservation copy of born-digital or digitally acquired content. As part of the licensing process, the library should procure permission from the content provider to retain a locally held physical copy sourced from the streaming file. After the e-resources unit completes the order, payment, and receipt processes, there will be a handoff to the unit with the staffing and equipment needed for burning DVDs, likely to be the same staff that would handle ripping from DVDs. In addition to ensuring that blank DVDs and jewel cases are readily at hand, printing labels for the item is another consideration to address. Once the burned DVD is available, it can go to the physical materials staff who will process and catalog the item as they would for purchased DVDs being added to the library collection. Concurrently, the e-resources unit can carry out their processing and access workflows.

Renewing Orders for Streaming Titles

Short-term licenses for streaming video are naturally predisposed to be treated as firm orders. With that said, some titles may be needed year-after-year; for example, a film that is required viewing for a recurring course that is core to a subject's curriculum. In these cases, an argument can be made for treating the title as a serial or subscription order, to be invoiced annually. Ultimately, this decision will come down to local practice. Librarians at the University of Northern Colorado grappled with this question while developing streaming workflows within their technical services unit (Leffler, Hayden, and Enoch 2017). While ultimately choosing to treat renewable titles as firm orders, similar to monographs, many within the library expressed the opinion that they were serials, by nature of being renewable, and should be treated as subscription orders.

Regardless of how the library opts to treat renewable streaming video orders, there are several considerations to address for titles or collections that will be renewed on a continuing basis. Firstly, even though the order is treated locally as a subscription, the content provider may not be positioned to offer renewal notices or annual invoicing. This should be established at the outset when placing the order, that the library intends to renew the resource annually. In some cases, a provider may not be able to offer this option, as they may not permanently retain the rights to continue to lease a particular title or collection.

Ideally, the streaming video content provider will offer the library advance notice when a termed lease is about to expire, but this is not always so, in which case the onus falls upon the library to track term expiration dates. This can be addressed by leveraging the functionality of a library's electronic resource management system (ERMS) to track expiration dates and by setting up tickler notifications through email. In the case of the University of North Colorado, library staff implemented ticklers for sixty days in advance of the expiration date, after difficult experiences relying on spreadsheets for tracking or vendors to provide notifications (Leffler, Hayden, and Enoch 2017, 104). A similar workflow of tracking, expiration notification, and renewal inquiry for short-term licensed streaming video was implemented at Santa Clara University using a mail merge within Google Sheets (Van Gaasbeck 2021). Other collaborative workflow management tools such as Trello or Microsoft Planner could also be used for tracking and expiration notification of short-term streaming licenses. Once a term expiration date is within the renewal period, the original requestor of the streaming video can be contacted for a decision on whether to renew for another term, potentially kicking off the ordering workflow process once again: the pricing is re-negotiated, a new license agreement may need signing, with ordering and invoicing to follow and a new expiration date to be tracked.

While the process for renewing streaming video orders isn't too dissimilar from other continuing resources, it often can be disjointed and entails more manual actions. For example, e-journals are typically subscribed to for the calendar year, resulting in a straightforward January 1 to December 31 term. Even database subscriptions from major vendors can be coordinated to have synchronous start dates so that the renewals can be grouped together. On the other hand, streaming video requests may be received throughout the year and ordered ad hoc, resulting in a continuous stream of expiration dates to potentially track. Another factor to consider is whether a film is locally managed or not, as the arrival of the term expiration will also require removing the video file from the local streaming platform, deleting the file, and removing access points from discovery tools. These considerations, which will be explored further in the next chapter, require a more hands-on approach for the management of streaming video, fundamentally different from other electronic resource formats that are more malleable to workflow regulation.

Conclusion

Ordering of streaming video can take a number of pathways, depending on the acquisition model being pursued. While the same basics of vendor communication and payment will apply across different use cases, additional factors such as the method of access and potential of bundling with physical formats can necessitate specialized workflows for receipt of content. Additionally, for many titles perpetual access will not be an option, requiring libraries to renew orders for streaming content in the same way they would other serial or continuing e-resources. More often than not, the acquisition models for streaming video favor access over ownership, a common frustration for libraries looking to build collections.

Streaming Vignette—University of Connecticut

Name of the library: University of Connecticut
Reporting librarians and titles: Jane Strudwick, Head of Acquisitions and Discovery; Merlita Taitague, Acquisitions and Electronic Resources Support Assistant
Type of library: Public land-grant R1 research institution, located in Storrs, Connecticut
User community: Full-time equivalent (FTE) of approximately 33,000 students
Total streaming budget: Approximately $75,000 annually in continuing funds

The University of Connecticut (UConn) has significantly grown its streaming video collection starting from the time Strudwick began working at the institution, in 2018, to the present. From relatively small beginnings, the growth of the streaming video collection has been managed in a strategic and conscious manner, with staff roles in the Acquisitions and Discovery department expanding to meet increased demand for streaming.

Collections-wise, UConn holds content from Kanopy, Swank, Docuseek, Alexander Street, Films on Demand, and Good Docs. Additionally, they work with a number of small and independent distributors to license individual films, many of which are locally hosted. UConn has an e-preferred policy for all formats; for video, streaming is the default, with DVD acquired when a streaming version cannot be sourced or licensed. The collection building is largely responsive, based on faculty requests that are made through a library-managed form and approved by subject specialists. Course reserve requests also come in directly and are handled by Taitague. Perpetual purchases are occasionally made with one-time funds. These are viewed as strategic additions to the collection, such as the Docuseek2 library of documentary titles that are highly supportive of research and instruction. Overall though, this "just-in-case" approach to collection building is difficult with the streaming video format because the majority of available content is licensed through short-term access models.

Licensing workflows are largely split between Strudwick and the e-resources librarian, who work on negotiating terms and pricing for content from the larger providers. For smaller distributors, Taitague takes a lead role in sourcing content, making inquiries about preferred terms, and verifying that the offered terms are in line with the practice for the larger providers, before proceeding with the acquisition.

UConn utilizes Kanopy's Smart PDA model for content from the provider, employing a highly curated and iterative approach. This approach was begun in 2022; prior to that, UConn had been pursuing one-off licenses for Kanopy titles. In the initial phase of the Smart PDA plan, UConn asked Kanopy to limit the available content pool to titles that had been previously licensed in the past three years. After an initial six-month period of assessing what titles were being triggered, UConn felt comfortable in expanding the pool of availability. With input from their Kanopy sales rep, content from a carefully selected "top five" group of suppliers was added. Through this measured approach to opening up content, title triggers for the plan have remained within the allocated budget. A potential next phase of the plan would be consulting with subject specialists to add further content and suppliers to the pool that would fall more under the "just-in-case" category.

For streaming video content that is locally hosted, files are uploaded to UConn's Kaltura instance, which is maintained by the university's IT department. Kaltura has been in use at UConn since 2019; prior to that, hosted content went through an additional migration, from a locally hosted server to the IT department's Mediasite instance. Over the course of multiple migrations, some locally held titles needed to be re-licensed, due to file format incompatibilities across the platforms. This experience served to highlight the tenuousness of life of file licenses, as media player compatibility and format obsolescence can raise issues with streaming in a vein similar to that of physical formats like VHS. Taitague handles digital asset management, downloading files sent from providers and then uploading video files and caption files when available. Receiving caption files can be a challenge with smaller providers; when negotiating these acquisitions, Taitague takes an advocacy approach in explaining why accessibility is necessary for serving all users. On occasion, UConn has not pursued acquisitions for providers who could not meet accessibility expectations; in one memorable example, the distributor of a documentary about hearing-impaired individuals would not provide captions for their video.

UConn uses Alma for their library services platform (LSP) with a Primo discovery layer; records are added for all streaming video titles. For short-term licensed titles, backend management is assisted through Outlook calendar reminders, so records for expiring titles can be removed. Since content is periodically prone to being pulled from collections, records for titles from Swank and Alexander Street include notes for faculty members to notify the library if they are intending to use a particular title for a course.

Regarding ongoing challenges for managing streaming video collections, Strudwick and Taitague cite the wide expanse of different acquisition models and license terms. For the Kanopy platform alone, some content may be part of the Smart PDA plan, some titles might be separately treated as an individual short-term license, and others may be perpetually licensed. This presents both an access and cataloging challenge in terms of managing title records. Taitague is currently working on cleaning up records in Alma to better reflect whether Kanopy titles are part of the PDA plan or licensed outright, adding end dates for short-term licenses as needed.

UConn Library, Collection Development Program, Policies, and Guidelines: https://lib.uconn.edu/about/welcome/policies-2/collection-development-program-policies-and-guidelines/

References

Koennecke, Jesse, Susan Marcin, and Matthew Pavlick. 2014. "The Devil Is in the Details: Managing the Growth of Streaming Media in Library Collections." *Proceedings of the Charleston Library Conference*. https://doi.org/10.5703/1288284315646.

Leffler, Jennifer, Jessica Hayden, and Todd Enoch. 2017. "Juggling a New Format with Existing Tools: Incorporating Streaming Video into Technical Services Workflows." *The Serials Librarian* 72 (1-4): 102–4. https://doi.org/10.1080/0361526X.2017.1284498.

Lowe, Randall A., Rick Davis, Leigh Ann DePope, Deborah K. Li, Robert K. Reeves, and Cindy Steinhoff. 2020. "Managing Streaming Video: The Experiences of Six Academic Libraries." *Journal of Electronic Resources Librarianship* 32 (2): 119–26. https://doi.org/10.1080/1941126X.2020.1739831.

Roll, Ann. 2019. "Report of the Session 'The State of Streaming Video in Academic Libraries,' Co-sponsored by the ALCTS Collection Management Section (CMS) Collection Development Librarians of Academic Libraries Interest Group and the Film and Media Round Table. American Library Association Midwinter Meeting, Seattle, January 2019." *Technical Services Quarterly* 36 (4): 408–10. https://doi.org/10.1080/07317131.2019.1664109.

Streeter, Abigail, Kimberly Kennedy, and Jamie Saragossi. 2023. "Streamlining Streaming Videos: Moving Towards an Effective Streaming Video Workflow in Alma/Primo." *Journal of Electronic Resources Librarianship* 35 (3): 171–81. https://doi.org/10.1080/1941126X.2023.2224663.

Van Gaasbeck, Kalvin. 2021. "Streaming Video Licensing Workflow Using Google Sheets." Presentation at NASIG, Virtual, May 19, 2021.

6 Local Hosting Challenges

Introduction

While a large majority of streaming video content is accessed through third-party vendor platforms, libraries are increasingly acquiring digital files directly from distributors for the purposes of self-hosting or local hosting. In many cases, this is done because the distributor lacks the infrastructure or platform to stream their content directly to viewers. This is typically seen with smaller companies and independent filmmakers, who retain the distribution rights to their content instead of partnering with a larger distributor. There are also larger streaming vendors, such as Kino Lorber, Video Data Bank, and Women Make Movies, who include files for local hosting among their acquisition options.

Local hosting of streaming files may be difficult if not outright unfeasible for many libraries. The primary reason is that a digital site license (DSL) to host a streaming video file may range from $150 to $500, with some titles costing even more than that. It may be prohibitively expensive for many libraries to spend these amounts on individual titles, in some cases only for a short-term license or possibly just a single year. Another potential barrier to local hosting is the infrastructure that needs to be in place, in terms of maintaining or subscribing to a content management platform that may entail monetary costs for upkeep or third-party hosting fees, as well as staff bandwidth to administer.

Despite these barriers, libraries are increasingly taking on responsibility for hosting locally held streaming video files. A 2021 *Choice* white paper published the results of a survey that found 36 percent of respondents were locally hosting licensed files at their libraries (Tanasse 2021, 8). One year later, a subsequent survey of streaming video practice in libraries found that nearly 48 percent of respondents were locally hosting content (Austin and Thornton 2022, 250). Semantics can be tricky in this area; the 2021 survey question described the hosting as being "on a local server" whereas the 2022 survey used the term "media management system." While local hosting is the commonly used term, libraries are increasingly more likely to use third-party platforms for this purpose over internal servers. "Locally managed" collections may be a more apt term, although "local hosting" is still useful as shorthand for the libraries holding such collections; this chapter will outline the various challenges that these local collections entail.

Local Hosting Options

The initial question for libraries acquiring streaming video files to locally host is where to put the files so that authorized users can view them? The answer can range from hosting on an internal server to third-party hosting solutions to utilizing file-sharing applications or content management systems. The library's particular needs for local hosting will be determined by a number of factors, including budget, size, number of files to be hosted, and any additional viewing features they may wish to utilize. Depending on the library, it is possible that pre-existing software applications or systems already in use at the library can be leveraged for the purpose of hosting streaming video. These could include cloud-based educational video platforms used for virtual classrooms and lecture capture, or systems used for course reserves. Some of the "quick and dirty" solutions will be briefly discussed, before evaluating more robust options. As is always the case with licensed materials, check the specific terms of the agreement, as they may prescribe requirements for security and authentication that will potentially rule out some options.

File Storage Applications

If a streaming file needs to be shared with a small number of individuals, for instance, students enrolled in a particular class, one of the most basic solutions could be utilizing cloud-based file storage and sharing applications such as Google Drive or Microsoft OneDrive. The file would be uploaded to a specific user account, and then view permissions can be controlled on an individual basis. The advantage of this method is that it is low or no cost, leveraging either freely available applications or applications already in use within the library. The limitations, however, are many: file sizes are often capped, the media player will be embedded in a browser with limited options, and granting permissions on an individual basis can be onerous. This method of handling local files is not advisable apart from special use cases where the local file only needs to be shared with limited individuals.

Free Hosting Sites

Popular streaming media sites such as YouTube and Vimeo offer users the ability to create free accounts and upload their own videos. Once uploaded, privacy settings and permissions can be modified so that only select individuals can view the content. Again, this type of solution would have limited application for local files acquired by a library and would not be advisable beyond use for a specific class with a small enrollment. As with file storage applications, the same concerns of file size caps will apply. Vimeo does offer an enterprise version of their hosting platform tailored to educational institutions. With that said, these sites are largely intended for locally created content and less appropriate for licensed material.

Course Reserve Platforms

Another option for sharing local streaming videos may be available through a library's course reserve system, often referred to as learning management systems (LMS). Some LMS, like Canvas or Moodle, offer the option of uploading media files into course pages. However, these uploads are typically subject to file size limitations that restrict files with longer running times or higher playback quality. As a result, the more common use cases for embedding video for course reserves will involve taking the embed code for an already hosted video on a site like YouTube and inserting it into the course page within the LMS. If the library uses a third-party media hosting platform like Kaltura or Panopto, integration with the LMS allows the embedding of already hosted video for courses.

An early case study in local hosting at the University of North Carolina at Greensboro (UNCG) used a combination of methods for hosting streaming video files and making them accessible to users. YouTube was initially considered but soon ruled out; the ease with which links could be shared and videos downloaded made it incompatible with license terms for copyrighted materials. It was determined that streaming files could be uploaded to Google Drive; UNCG was utilizing Google Apps for Education, which allowed for access to be limited within the institution. To further regulate access, the video links would then be embedded into the Blackboard LMS by the instructors putting the titles on course reserve (Cross, Fischer, and Rothermel 2014, 157). Creative application of already existing tools has been common for libraries grappling with local hosting challenges, particularly those who were early adopters of the format.

Local Servers

When local management of individually held collections of streaming files was relatively new, libraries frequently resorted to internally managed servers for hosting. These homegrown solutions often featured locally developed user interfaces on the front-end, created utilizing open-source software. One early adopter was Concordia University, who piloted local hosting of streaming audio content on password-protected library-dedicated servers in 2008–2010 (Doi, Mason, and Wiercinski 2011), before moving to the inclusion of locally hosted streaming audio content in 2014, then expanding their streaming video collection with content hosted on third-party vendor platforms (Smith 2022, 277–78). When the libraries at Brigham Young University began investigating options for streaming video to patrons in 2006, they ended up partnering with information technology staff to create Byugle—a locally networked solution—in part because the available vendor-provided streaming services at the time were prohibitively expensive (Schroeder and Williamsen 2011, 91–3). At Texas State University (TSU), steady growth in demand for resources to support distance learning coincided with a transition from locally hosted password-protected servers for video streaming to a local instance of third-party streaming platform software in 2015 (Towery, Price, and Cowen 2019, 4–5).

Frequently, these early hosting solutions were borne out of necessity; video content was being created locally and institutions needed a place to store it. Gradually, third-party software solutions were embraced as libraries began to acquire more licensed streaming files and more hosting providers entered the market. A 2016 article by farrelly noted this shift, outlining and listing several of the third-party hosting, or "turnkey" solutions appearing in the academic marketplace over the preceding years (farrelly 2016/17, 178–9). One of the earliest published studies of locally managed streaming video—at Borough of Manhattan Community College—utilized a third-party or turnkey solution, the now defunct Axonix SuperView platform (Eng and Hernandez 2006, 218). While the term turnkey solutions is no longer generally used to refer to media hosting platforms, the current tools on the market all similarly function to perform common tasks and each has unique advantages and drawbacks and is likely to require customizations for local implementations. The next section will provide an overview of some of the more prominent hosting solutions currently used by libraries.

Third-Party Platforms

The previously discussed hosting options have limited applicability, mostly confined to sharing video content for individual courses. For libraries looking to build and maintain collections of streaming video, more robust platform options should be sought out to maximize the visibility and usage of the collection. As noted, a number of third-party options are on the market that libraries can utilize. These types of hosting solutions fall under the software-as-a-service (SaaS) licensing and delivery model, where the provider offers a cloud-based instance of the application. The library licenses the service and pays a subscription cost, and users can access and administer the platform through a web browser. As with most other enterprise solutions, costs can vary widely, with many options being prohibitively expensive for libraries. While this chapter will not delve into the process of licensing a streaming video-hosting service, many third-party providers offer tiered pricing models based on factors such as institutional size or the amount of content that is to be hosted on the platform. The following summaries outline some of the more common third-party media platforms being implemented by libraries at the time of this writing. Functionality varies widely, and some products are primarily intended for creating streaming video, such as pre-recorded course lectures. In no way do the authors endorse one platform over the other, though, in the interests of full disclosure, two of the products outlined (Aviary and Panopto) are currently in use at the authors' institution.

Aviary

https://www.aviaryplatform.com/

Aviary is a relatively new media streaming platform, having launched in early 2019. It offers cloud-based hosting services, with subscription tiers incremented by a maximum number of hosted resources. Pay-as-you-go plans that charge for each resource added are also offered. Access permissions can be managed at a granular level by individual resources and supported access management methods include IP addresses, email domains, as well as support for single sign-on (SSO) authentication. While access can be restricted to authorized users only, Aviary customers have the option to make their content and metadata discoverable through the public Aviary portal, which includes published resources from Aviary instances at other institutions. Aviary's functionality lends itself to curated digital archive collections; prominent examples of institutional implementations include the Hip Hop in Queens collection of the Queens Public Library, Indiana University's Bicentennial Oral History Project, and the Fortunoff Video Archive for Holocaust Testimonies.

Supported file formats: MP4 recommended for video
Limits on file upload size: Up to 50 GB supported, recommended 5 GB or less
Accessibility functionality: Upload of caption files supported Video Text Tracks (VTT format), supports integration with captioning applications via application programming interface (API)
Metadata: Descriptive metadata elements based on Dublin Core schema, allow for unlimited custom fields, thumbnail images supported
LMS integration: Not available
Usage analytics: Custom analytics dashboard, includes individual plays, unique viewers, minutes viewed, and user engagement statistics, for example, average completion rate by video

Echo360

https://echo360.com/education/

Echo360 is widely used in education for its video creation and lecture capture functionality, allowing instructors to create and share recordings with students. Content creators can edit videos within the platform and Echo360 integrates with the major LMS platforms. It can also be used as a content management system for maintaining collections of externally licensed videos. Videos are automatically transcribed with automatic speech recognition and transcripts can be manually edited.

Supported file formats: MP4, MOV, WMV, AVI, FLV, others
Limits on file upload size: 20 GB
Accessibility functionality: Screen reader support, automated captioning, upload of caption files supported (SubRip Subtitle (SRT) and VTT formats), audio description tracks partially supported, Web Content Accessibility

Guidelines (WCAG) 2.1 Level AA compliance (supports or at least partially supports all criteria), Voluntary Product Accessibility Template (VPAT) made through third-party audit available on the website

Metadata: Name, Description, and Tags for individual media files

LMS integration: Blackboard, Canvas, D2L Brightspace, Moodle, Sakai

Usage analytics: Custom analytics dashboard which includes individual plays, unique viewers, total and average minutes viewed, and user engagement statistics such as timestamped heat mapping overlay

Kaltura

https://corp.kaltura.com/education-video-platform/

For educational institutions, Kaltura is among the most widely adopted media streaming platforms. Current institutional customers include Harvard University, UCLA, and the University of Virginia, among others. Their services allow clients to not only upload, host, and share existing video files but also to create and livestream video. These types of applications have wide utility in an educational context, as instructors can pre-record lectures or hold courses in a virtual classroom. Students can upload their own video assignments as well. Accessibility features include automatic captioning. In addition to a cloud-based SaaS implementation of Kaltura, institutions have the option to self-host an implementation of Kaltura on their own servers, either through a licensed on-premises edition or freely, as the platform code is open source and supported by a user community.

Supported file formats: MP4, MOV, WMV, AVI, AVCHD, FLV, others

Limits on file upload size: Unlimited

Accessibility functionality: Screen reader support (all major screen readers), keyboard shortcuts (partial support), automated captioning, human-generated captions available upon request for a fee, upload of caption files supported (SRT, VTT, and DXFP formats), audio description tracks supported, Section 508 compliance, WCAG 2.1 AA compliance (supports or at least partially supports all criteria), VPAT available on website

Metadata: In addition to basic metadata fields (Name, Description, and Tags), it supports custom metadata schema

LMS integration: Blackboard, Canvas, D2L Brightspace, Moodle, Sakai

Usage analytics: Custom analytics dashboard which includes individual plays, unique viewers, minutes viewed, and user engagement statistics such as average completion rate by video

Mediasite

https://mediasite.com/education/

Mediasite offers a host of functionality in support of virtual education, including livestreaming to the virtual classroom, lecture capture, and the ability for students to record video assignments. Mediasite supports SSO for authenticating users and integrates with all major LMS platforms. Viewing permissions can be set for all authorized users within an organization or at a granular level for specified individuals and groups. Accessibility is supported by allowing users to manually upload caption files for videos; automated transcription can also be applied to videos. Mediasite is currently utilized at a number of institutions of higher education including Villanova, Ohio State University, and Kansas State University.

Supported file formats: MP4, MOV, AVI, FLV
Limits on file upload size: Unlimited
Accessibility functionality: Screen reader support, keyboard shortcuts, automated captioning, upload of caption files supported (SRT, SAMI, and DXFP formats), WCAG 2.1 AA compliance (supports or at least partially supports all criteria), VPAT available on website
Metadata: Name, Description, Presenter/Author, and Tags for individual media files
LMS integration: Blackboard, Canvas, D2L Brightspace, Moodle, Sakai
Usage analytics: Custom analytics dashboard which includes individual plays, unique viewers, average minutes viewed, and user engagement statistics such as timestamped heat mapping for individual videos

Panopto

https://www.panopto.com/panopto-for-education/

Panopto is another video-hosting platform that has been widely adopted across numerous educational institutions, including Johns Hopkins University, Cornell University, and the University of Michigan. Much like Kaltura, the platform allows for uploads of locally created and licensed content, as well as options to support virtual learning, such as livestreaming, video conferencing, and lecture capture. Integration with major LMS providers is another feature of the platform. With regard to accessibility, Panopto offers a range of captioning options. Uploaded videos are automatically transcribed; customers can also request for Panopto to provide Section 508-compliant human-transcribed captions on a per- title basis for an additional cost. Additionally, Panopto partners with several third-party captioning services. Audio descriptions are supported as well.

Supported file formats: MP4, MOV, WMV, AVI, FLV, others
Limits on file upload size: Unlimited
Accessibility functionality: Screen reader support, keyboard shortcuts, automated captioning, upload of caption files supported (SRT, VTT, ASHX,

and DXFP formats), human-generated captions available upon request for a fee, audio description tracks supported, Section 508 compliance, WCAG 2.1 AA compliance (supports or at least partially supports all criteria), VPAT available on website

Metadata: Name, Description, and Tags for individual media files

LMS integration: Blackboard, Canvas, D2L Brightspace, Moodle, Sakai, Cornerstone, Bridge, other Learning Tools Interoperability (LTI)-compatible LMS

Usage analytics: Custom analytics dashboard which includes individual plays, unique viewers, minutes viewed, and user engagement statistics such as average completion rate by video, supports integration with Google Analytics for customized reporting

Local Hosting through Streaming Content Distributors

A number of the larger distributors who license streaming video content also offer customers the option to host locally maintained collections through their platforms. In many cases, this is provided free of charge as an additional feature of the platform. For libraries that lack the infrastructure to host on local servers or the budget to pursue an enterprise solution CMS from the previous section, this is a potential option, provided of course that they are already licensing content on the vendor's platform. A possible bonus—in terms of the user experience—is that local collections may be made discoverable through the platform along with other videos from the vendor and that the interface will already be familiar to users. Libraries should confirm specifics before devoting energy to triaging collections on vendor platforms. For example, are there limits to how many individual titles or files can be hosted? Can content licensed from external rightsholders be hosted, or is it limited to locally created content (e.g., lectures, class recordings)? At a minimum, it is most likely that the library will need to retain subscriptions from the vendor in order to maintain the service, but it should also be clarified if there is a certain threshold of annual spend on the vendor's content that is required. Below is a listing of streaming distributors who currently offer local hosting services. As a note, although Kanopy had in the past provided a hosting service—and does maintain it for legacy customers—they do not currently offer hosting, therefore they are not included.

Alexander Street Media Hosting Service

https://alexanderstreet.com/products/media-hosting-service

Alexander Street, now part of Clarivate, is one of the most long-established and widely used distributors of streaming video content. Its flagship collection, Academic Video Online, contains over 80,000 multidisciplinary titles and is considered a core resource for many libraries. The Media Hosting Service is offered to Academic Video

Online subscribing libraries upon request and free of additional charges. Libraries can upload their locally licensed video files through the platform admin portal. Custom metadata can be included and the locally maintained content will be discoverable on the platform alongside other licensed Alexander Street collections and titles. The service allows users to upload an unlimited number of files, though availability is contingent on continued maintenance of the Academic Video Online subscription. Upon upload of a file, there is a free-text metadata field for a copyright statement, where the rightsholder can be noted. An expiration date can also be entered, after which the file will be automatically deleted from the platform.

Supported file formats: MP4, MOV, WMV, AVI, FLV
Limits on file upload size: 4 GB
Accessibility functionality: Screen reader support, keyboard shortcuts, upload of caption files (SRT supported), Section 508 compliance, WCAG 2.1 AA compliance (supports or at least partially supports most criteria), VPAT available on website
Metadata: Pre-defined metadata fields for file uploads; required fields include Title, Abstract, Field of Study, Publisher, Publication Place, and Publication Year; optional fields include Director, Local Identifier, Series information, and Copyright Statement; thumbnail images supported
LMS integration: Blackboard, Canvas, Brightspace/D2L, Moodle
Usage analytics: COUNTER R4 and R5 reports; custom analytics dashboard with playback metrics, user engagement measures (e.g., number of videos cited, shared, embeds), accessibility measures (number of videos played with onscreen transcripts)

Infobase

https://infobaseadmin.zendesk.com/hc/en-us/articles/360036166013-Infobase-Hosted-Custom-Content

Primarily known as a distributor of streaming video content for collections such as Films on Demand, Infobase also offers a service, Custom Content, allowing libraries to maintain collections of their licensed or created streaming videos to locally stream on the Infobase platform. Libraries have the option of self-hosting on their own internal server or they can upload videos to the Infobase cloud. The pricing model is tiered, with levels structured according to the number of videos being hosted. The initial ten videos are included for free to Films on Demand subscribers. Extra paid options include DVD-to-digital conversion and segmenting of media files. When uploading new videos, users are required to check a box certifying that the proper license has been obtained for the content to be hosted. The platform also allows for setting expiration dates on videos.

Supported file formats: MP4, MOV, WMV, AVI, FLV, others
Limits on file upload size: 5 GB
Accessibility functionality: Screen reader support, keyboard shortcuts, automated captioning, upload of caption files supported (SRT and DXFP formats), human-generated captions available upon request for a fee, audio description tracks supported, Section 508 compliance, WCAG 2.1 AA compliance (supports or at least partially supports most criteria), and VPAT available on website
Metadata: Pre-defined metadata fields for file uploads; required fields include Title, Description, Producer, and Grade Levels; optional fields include Keywords/Tags, Series information, and Copyright Year; freely exportable machine-readable cataloging (MARC) records available via admin portal; thumbnail images supported
LMS integration: Blackboard, Canvas, Brightspace/D2L, Moodle, other LTI-compatible LMS
Usage analytics: Non-COUNTER reporting measures Views, Searches, and Clicks on the platform; COUNTER R5 reports are available for Films On Demand but do not include metrics for Custom Content

McIntyre Media

https://www.mcintyre.ca/streaming-and-hosting-services

McIntyre is a Canada-based video content distributor that primarily serves educational institutions in Canada and is the regional distributor for Films on Demand in addition to licensing individual titles and collections. They also provide hosting services for the management of local streaming video collections. Their basic service allows administrators to add metadata to content and view usage statistics. Prior to uploading files, users are asked to confirm they have secured the rights to host the content. Expiration dates can be set for files or be marked as "perpetual" when access is licensed for life of file. Additional services offered include digitization, closed-captioning, and transcript creation.

Supported file formats: MP4 format required for uploaded videos
Limits on file upload size: Unlimited
Accessibility functionality: Upload of caption files supported (SRT and VTT formats), human-generated captions available upon request for a fee
Metadata: Pre-defined fields include Title (required), Description, Series information, Producer, Release Year, Subjects, Grades; thumbnail images supported
LMS integration: Embed code provided for integrating content into LTI-compatible LMS
Usage analytics: Admin portal provides customizable usage reports

Digital Repositories

Another potential solution for the management of locally held streaming video could be found in leveraging the use of an institutional repository (IR), such as bepress Digital Commons, Islandora, or DSpace. IRs may already be in use by the library and are frequently utilized for archiving scholarly output by faculty at an institution, as well as being used as platforms for publishing content. Streaming media file formats are typically supported, and some IR platforms, like Islandora and DSpace, utilize freely available open-source software. Some library service platforms (LSP) also offer digital content management functionality. Ex Libris's Alma LSP, for example, offers its Digital Resources solution as a tool for depositing electronic content and managing controlled access. Prospective use cases for this management solution could be digital course reserves or controlled digital lending (CDL). As the functionality of these systems expands to include digital content management, libraries may turn to them for streaming video hosting.

Selecting a Hosting Platform

Libraries will want to scale their choice of a third-party hosting platform to their specific use cases for locally managed streaming video collections. Obviously, cost will be a major determining factor to consider, as will the size of the collection being managed. As noted, many of the options on the market offer staggered pricing tiers based on the number of titles being hosted (e.g., Tier 1 offers pricing for up to 100 hosted videos, Tier 2 pricing is for up to 1,000 hosted videos). Another consideration is how robust the functionality is compared to what is actually needed by the library. Some of the options covered are primarily intended for virtual classroom use cases and feature lecture capture and livestreaming functions that go well beyond basic content management. With that said, if the platform is already in use at the library's parent institution, it may be worth investigating if it can be leveraged to manage locally held streaming video collections. In doing so, the library's set of needs will be less than those of virtual instruction, and more oriented toward technical services areas such as access, metadata, and discoverability. Levels of customer service available from the provider are another consideration, as is how the platform complies with accessibility guidelines with support for captioning and assistive technology. Many of the same assessment principles used in streaming video collection development can be applied to the selection of a platform for hosting the library's locally held content.

A 2021 Ithaka S+R survey of streaming media strategies in academic libraries asked respondents to indicate the platforms they used for digitization and/or hosting local media content, with the option to choose multiple responses (Cooper, Ruediger, and Skinner 2022). A full two-thirds, or 66 percent, of libraries indicated that they self-hosted using tools such as LMS and internal servers/repositories, supporting the practice of libraries leveraging existing systems for content hosting. Services like

Kaltura (29 percent of respondents) and Panopto (17 percent of respondents) were also widely utilized (Cooper, Ruediger, and Skinner 2022, 17–18). Finally, hosting options from content distributors also had noteworthy uptake from libraries, with Alexander Street, Kanopy, and Infobase indicated as being used by respondents (Cooper, Ruediger, and Skinner 2022, 17–18).

Handling of Local Streaming Video Files

Video Resolution

Once a library has settled on a platform solution for hosting and streaming locally held video files, there are a number of concerns to address regarding how the files themselves must be handled. On the backend, a master copy of the file will need to be stored, possibly on a local network or server. This would be considered an archival copy of the file and may also be the same version that is uploaded for streaming on the library's local platform. In some cases, streaming video distributors may provide multiple versions of the same title, offered at different resolutions. Commonly, video files are provided at 480p and/or 720p resolution, respectively categorized as Standard Definition and High Definition. The resolution numbers correspond to the number of pixels in the image, with a greater number of pixels corresponding to a higher image quality. The resolution is actually made of two values, the pixel width by the pixel height (e.g., 640 x 480), but the shorthand reference is to the vertical pixel value or pixel height. So a resolution of 640 x 480 is referred to as 480p. Interestingly, the "p" does not stand for "pixel," but for "progressive scan," which is how the pixels enter into each new frame of video. Table 6.1 provides an outline of the common resolution types and their shorthand names and dimensions, based on a 16:9 aspect ratio, which aligns with the screen width and height for most modern computer monitors, smartphones, and televisions. This is a widescreen or rectangular aspect ratio, compared to the more square-like 4:3 aspect ratio that was common to film and television produced in the 1980s and 1990s.

Video files with 480p or 720p resolution will more than likely provide high enough image quality and clarity for libraries to stream to their users. Distributors may

Table 6.1 Video Resolution Specifications

Resolution Name	Pixel Size	Resolution Type
480p	720 × 480	Standard Definition (SD)
720p	1280 × 720	High Definition (HD)
1080p	1,920 × 1,080	Full HD
4K or 2160p	3,840 × 2,160	4K Video or Ultra HD

also provide lower-resolution files, for example in 360p, and these could also be sufficient for streaming to users. The lower-resolution videos will be less prone to buffering lag, which could be beneficial for users with slower internet connections. Lower-resolution videos also have smaller file sizes, which may be a consideration if there are limitations for individual file uploads on the library's streaming media platform.

If the distributor offers multiple versions of a streaming file, it can be a good practice for libraries to download and store all available versions, as multiple use cases may apply. For example, the library may want to designate one version as an archival or master copy for preservation purposes, and another as a usage copy for upload to the streaming platform. In this scenario, a high-resolution copy, for example, 720p or even 1080p, could serve as the master copy, whereas the lower-resolution version would be the derivative, or usage copy. The library could even choose to upload both Standard- and High-Definition versions of a title, to provide alternatives for users with slower connection speeds, or possibly those streaming through a mobile device who may not need the higher picture quality. Figure 6.1 shows how multiple-resolution copies can be stored.

Ultimately, a library's policy will be based on a combination of considerations. Are there file size limitations on their streaming platform? What are the expectations for picture quality? Do users have quality connectivity and on what devices will they stream the library's content? In many cases, distributors may only have a single version of the streaming file that is provided to libraries, so it's important for specifications such as resolution, format, and file size to be established up front prior to acquisition.

(\\storage.yale.edu\home) (I:) › Locally Hosted Content › Streaming media › Chinatown Rising

Name	Type	Size
Chinatown_Rising_360p.mp4	MP4 File	367,425 KB
Chinatown_Rising_540p.mp4	MP4 File	914,760 KB
Chinatown_Rising_720p.mp4	MP4 File	1,547,296 KB
Chinatown_Rising_1080p.mp4	MP4 File	3,122,157 KB
Chinatown_Rising_1080p_masterfile.mp4	MP4 File	9,802,926 KB
Chinatown_Rising_captions_Chinese.srt	SRT File	77 KB
Chinatown_Rising_captions_Chinese.vtt	VTT File	74 KB
Chinatown_Rising_captions_English.srt	SRT File	81 KB
Chinatown_Rising_captions_English.vtt	VTT File	77 KB
Chinatown_Rising_poster.png	PNG File	3,689 KB
Chinatown_Rising_thumb.png	PNG File	117 KB

Figure 6.1 Storage of streaming video with multiple-resolution options.

File Format

File format is another area where libraries need to take into account the potential constraints of their locally managed streaming platform. The most commonly used digital video format is the MPEG-4 or MP4 format; MP4 is an international standard and is supported across all the major streaming platforms. The other frequently encountered digital video format is MOV. This format was developed by Apple specifically for use with their QuickTime Media player. Currently, MOV is supported across other media players but not to the degree that MP4 is. For example, Windows Media Player will play MOV files, but VLC Player will not. MOV files may be favored by content creators due to the fact that they are easy to edit and have high picture quality with little to no compression. However, local hosting of MOV files may not be possible if the library's streaming platform does not offer support for the format, or if the file sizes exceed the platform's limitations for individual uploads. Although MP4 is an internationally defined standard under ISO/IEC 14496-14 the current Recommended Formats Statement of the Library of Congress categorizes MOV, encoded as V210, as a Preferred format for video files, while MP4, encoded as H.264, falls under Acceptable formats as a viewing proxy (Library of Congress 2022). Similarly, at the authors' institution, MOV is the preservation master format for video, while MP4 is considered an access copy format. The lack of compression for MOV makes it a better candidate for archival copies, while the smaller file sizes for MP4 make it more "portable" and better geared toward usage copies.

The best practice for libraries regarding file formats and file sizes will be to communicate known preferences or requirements to the vendor or distributor so that they can take steps to make any necessary format conversions to provide a usable file. Unfortunately, in some cases files will already be received before it is discovered that they are in an unsupported format or the file size is too large. When this occurs, the first step should be to communicate the issue to the vendor and request a replacement file. If the vendor does not have the ability to do so, or in cases where time constraints are an issue, the library can potentially create a usable copy or derivative file. If this right is not stated in the original license or agreement with the vendor, the library should make contact to request permission first. Once that is done, there are a number of ways to go about creating a derivative file. If there is a media center in the library, they may have specialized software, such as Adobe Pro, to convert or compress video files. There are also freely available applications—like Handbrake (https://handbrake.fr), an open-source transcoding program—which can be used to compress video files and convert formats, such as MOV to MP4. Another potential option, if the title is also physically owned on DVD, is to rip a version of the file that will adhere to the requirements of the library's streaming platform. While these options may be viable, they require permission from the vendor, software resources, and staff time, which places an additional onus on the library. The first and best course of action is to request a usable version of the file from the vendor.

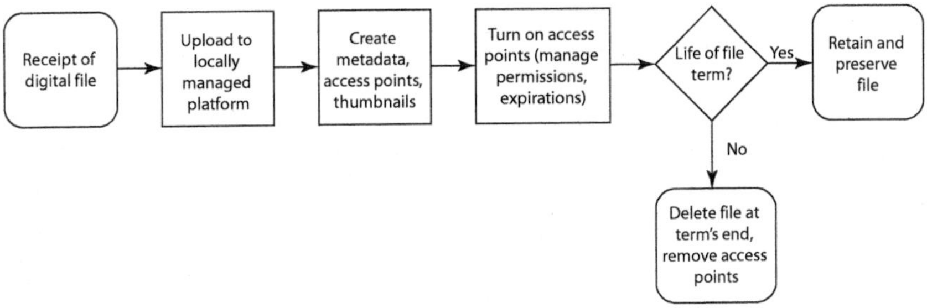

Figure 6.2 Workflow for locally managed streaming video.

Workflow for Local Management

Locally hosted streaming video files largely adhere to a general technical services workflow for receipt, access, and maintenance. Format-specific factors to consider will vary, depending on the use case and license term. Figure 6.2 outlines the general steps for local management, which will be discussed in greater detail.

Receipt and Upload of Files

The initial workflow steps of acquisition include both licensing and ordering and were covered in detail in Chapters 4 and 5, respectively. As noted in the ordering chapter, receipt of the files to be locally managed can occur in a few different ways. A distributor-provided download link is the most common method of delivery, but the files could potentially be sent on a physical device such as a USB drive, or they may be ripped, with permission, from a physically owned copy of the title.

The files will likely have to be saved locally prior to uploading to the streaming platform. Ideally, they will be saved to a closed network server where the master copies of streaming files will be housed and where they can be accessed by designated library staff. A good practice is to create a hierarchical file structure arranged by distributors, with subfolders for individual titles, as seen in Figure 6.3. Each title subfolder will then house not only the master copy of the streaming file but any supplemental files as well. These could include additional use copies of varying resolutions and file sizes, caption files for accessibility, thumbnail images created for the hosting platform, promotional images or documents supplied by the distributor, or additional video files such as deleted scenes, commonly included as "extras" on physical DVDs or Blu-ray discs. Utilizing standard naming conventions is another best practice for local management, as recognizable handles for files will assist the library staff and erase any guesswork of the default distributor file names, which can often be long and unwieldy.

The next step will be for library staff to upload files to the streaming platform. Unsurprisingly, the upload process can be quite time-consuming, particularly for feature-length titles encoded at high resolutions or in uncompressed file formats like MOV; individual file sizes can often be several gigabytes. Upload times may

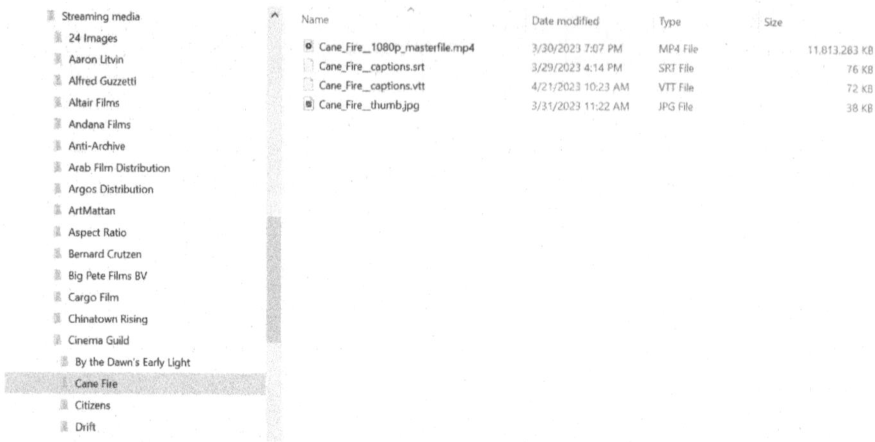

Figure 6.3 File organization for locally managed streaming video.

vary, depending on the bandwidth and capacity of the hosting platform being used. This may be less of an issue in the case of automated bulk uploads, but the more likely use case is that staff members will manually upload individual files as they are acquired. Although staff can multi-task while file uploads are occurring as background processes, checking on progress statuses while doing so can disrupt the flow of their other work. To facilitate this step of the workflow, a library could designate a standalone computer terminal for uploading files; particularly large files could be targeted for overnight uploads. If there are supplemental files to include with the video file—for example, a separate caption or transcript file—they are generally added after the initial upload completes.

Descriptive Metadata and Thumbnail Images

At the point of upload, or following the upload's completion, descriptive metadata for the title will also be added. The available metadata fields will vary by the hosting platform, but at the barest minimum, a title will be needed to distinguish the uploaded file. Additional information could include the name of the filmmaker or distributor, the distribution year, a description or summary of the film, subject headings/tags, and pertinent information from the license agreement, such as whether public performance rights (PPR) are available. It is important to note that the descriptive metadata being input directly to the hosting platform will only be discoverable to users who are already searching within the platform. Other metadata such as MARC records of locally managed streaming files for the library catalog will be discussed in the next chapter. Still, it is a good practice to populate at least some basic metadata at the platform level that will be user-facing.

In all likelihood, the most prominent user-facing aspect of the film will be a thumbnail image of the video itself. Many hosting platforms will automatically generate a thumbnail image after the upload completes, typically in the form of a screencap, taken from a random point during the video. Depending on the timestamp of the preview image, it may not be a representative frame from the video or may even be a blank screen. On popular streaming platforms like YouTube, the video thumbnail can be all-important to entice users to press play. That is not necessarily a motivating factor for library collections, but a thumbnail should at least give users a sense of what the video is. Depending on availability, a promotional image for the video could be used, such as a poster. Since not all titles will have readily available promotional images, a good practice is to create a screen capture of the video's title frame to be used as the thumbnail. This will provide a useful visual accompaniment and reinforcement of the title information.

Any type of screenshot tool can be used to capture and save a thumbnail image. Media editing software like Adobe Premiere Pro has the ability to isolate and save individual frames of a video. Simpler solutions also exist; for example, playing the video on a media player like VLC or Windows Media Player, pausing on the desired screen, and using the program's snapshot or save image functionality. An external

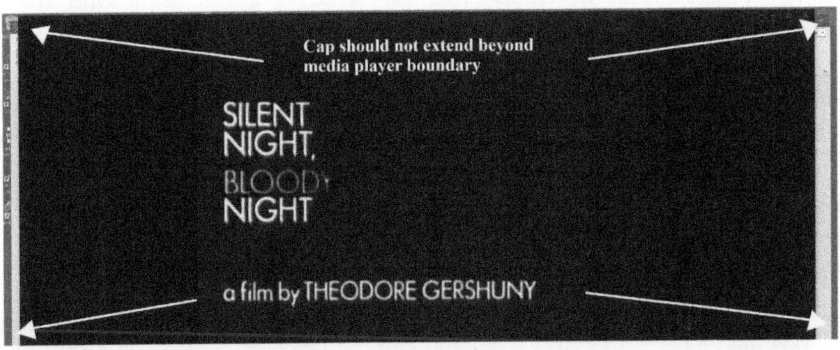

Figure 6.4 Screencap outside media player window.

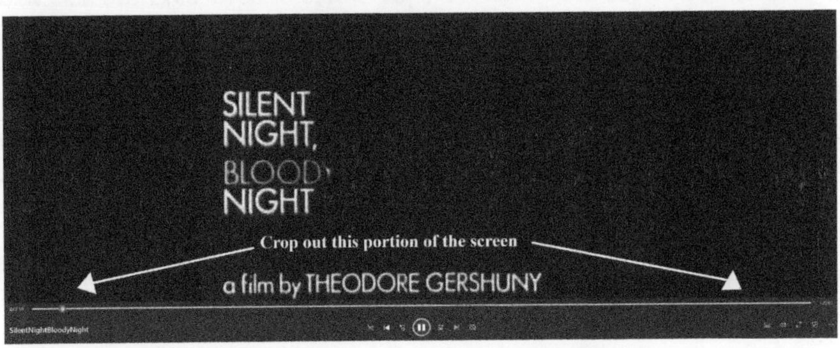

Figure 6.5 Screencap with uncropped media player control panel.

screengrab program, like the Windows Snipping Tool or free programs like Lightshot, could also be used in this way, though it is necessary to take the screengrab from within the display of the media player (Figure 6.4) and be sure to crop out anything external to the image such as the toolbar or control panel of the media player (Figure 6.5). Even simpler, your computer's OS may have hotkeys equipped for screen capture; for example, on most Windows computers pressing the Windows logo key + the PrtSc button will generate an image that is autosaved in the C:\Users\...\Pictures\Screenshots folder. However, a full-screen capture will likely require editing/cropping to get a desirable thumbnail image. For title screen thumbnails, it is ideal if the title is displayed against a solid background, making it easier to read. If the title screen displays over an image, it can be a judgement call as to what looks best as a thumbnail. The example in Figures 6.6 and 6.7 shows a title screen with and without a background image, respectively. While Figure 6.6 is more visual, Figure 6.7 presents the title more clearly, and could be preferable, particularly if thumbnails display in a small size.

Providing Access

Once a video file has been uploaded to a streaming platform and the necessary descriptive metadata has been added, access can be provided to users. When making locally held streaming video content available, there may be questions about to whom it should be made available and for how long. Depending on the hosting platform, permissions and access levels can be set. While some libraries may have some locally held video archive collections that they wish to make fully open to the public, licensed streaming video files are restricted to authorized users. Authentication can be mediated in the same way that vendor-hosted e-resources are, through methods such as library-specific IP address or SSO authentication. Some platforms allow videos to be shared with individual users or a URL link can be generated that allows viewing for any authorized viewer within the organization. If a video is being used for course reserves, it is common for that URL to be entered into the reserves LMS, so that only those enrolled in the course have access to it. In some cases, a user accessing one video on a platform can cross-search within the platform to discover other available videos uploaded to the platform. Depending on the platform, settings can be modified so these titles don't display as discoverable and are only accessible to users with the URL.

Another useful function to manage access that many streaming platforms provide is the ability to set limits on a video's availability. Within the platform's admin, a start and end date and time can be set at an individual title level. Figure 6.8 shows how this can be done on the Panopto platform. This can be useful for videos that have short-term licenses, videos being put on temporary digital reserve for a course, or "virtual screenings" of videos with limited availability windows, that is, twenty-four to forty-eight hours.

Figure 6.6 Thumbnail with title and background image.

Figure 6.7 Thumbnail with title only.

Figure 6.8 Availability settings on locally managed streaming video.

Retention, Renewal, or Removal

Issues of long-term access and retention will depend on the license terms negotiated for streaming video files, with removal and deletion of files sometimes a necessary step of the local management workflow. While many distributors license streaming video titles for the life of the file, short-term licenses are sometimes the only option made available. Typically, the licenses will explicitly state instructions for the file's removal. The file should not only be removed from the streaming platform but all locally held derivative files should be deleted as well, including those saved to any local network folders or servers. This poses a management challenge for libraries, as they need to track the terms for short-term licensed streaming files, in order to be able to delete the files at the expiration of the term. In cases where libraries may need to renew a short-term license, it is a best practice to do this ahead of the term expiring. In that way, the library can continue to use the locally held file without going through the workflow steps of file receipt and upload once again. When renewing these licenses, it should be communicated to the content provider that the file is already locally held. Proactively flagging short-term licensed titles intended for renewal can also build in time to make the process run more smoothly.

Even when a locally held title is licensed for the life of the file, issues such as format obsolescence can cloud the long-term preservation picture. As some have noted, if file formats become obsolete or incompatible with current standards for devices and browsers, it could mirror a cyclical need to re-acquire content anew, similar to physical media formats advancing from VHS to DVD (Laskowski and Teper 2014, 61). Ideally, the library will negotiate license agreements that secure the right to convert the original master file into new formats for access copies. System migrations onto new streaming platforms may also introduce the need for updating access copies to ensure the long-term usability of a locally managed streaming video collection.

Conclusion

Managing locally held collections of streaming video files presents a myriad of challenges, as it puts the library in a more prominent role of directly providing content to users. Selecting and maintaining a hosting platform, handling large files, and regulating access for authorized users are all factors to consider. While the practice of managing local streaming collections puts additional onus upon library staff, it is also an opportunity for libraries to retain greater intellectual control over their licensed content than they would have with content that is licensed by and streamed over a third-party vendor's platform. Local management of streaming video allows libraries to build collections that can potentially be as unique and diverse as physical media collections while providing the type of on-demand access that users have come to expect.

Streaming Vignette—University of Michigan-Dearborn

Name of the library: University of Michigan-Dearborn (Mardigian Library)
Reporting librarian and title: Timothy Streasick, Electronic Resources Librarian
Type of library: Public doctoral-granting university in Dearborn, Michigan
User community: Full-time equivalent (FTE) of 8,223 students; many are commuter students and nearly half of the student body are first-generation college students
Total streaming budget: $15,000 annually

The University of Michigan-Dearborn (UM-Dearborn) is a regional university located in the Detroit metropolitan area and operating as part of the University of Michigan in Ann Arbor, along with the University of Michigan-Flint. The UM-Dearborn library has its own collections budget, although some content is shared among all the University of Michigan libraries, or made available through the Michigan eLibrary (MeL), a state library consortium that provides resources for libraries of all types, including public and K-12 school libraries.

UM-Dearborn's streaming video collection includes content from Kanopy, Swank, and Alexander Street, as well as a small number of DSL titles that are locally hosted. Kanopy was moved from an unmediated PDA to a mediated plan in the 2018/2019 fiscal year after a usage analysis found that many titles saw little usage after being triggered. The mediated plan allowed for Kanopy to be budgeted for as a stable line item from year to year. By the time COVID-19 forced a move to all remote learning in 2020, the process of mediating and budgeting for streaming was running quite smoothly. The streaming video budget is managed by Streasick, who also has selection duties for streaming. The streaming budget is allocated from interdisciplinary funds, although on occasion a video request will come in from a subject selector.

Approximately 90 percent of UM-Dearborn's licensed streaming video content is in response to curricular requests from instructors. Occasional one-time or grant money has allowed Streasick to do more permanent collection building by acquiring life of file DSLs. Selection of titles is also largely based on curricular needs—films that have been used in the past for courses or ones that likely will be. Locally hosted streaming files are uploaded to the University of Michigan's shared instance of Kaltura. Streasick oversees the various aspects of UM-Dearborn's locally managed streaming video. Separate caption files are uploaded and tested to ensure that they sync correctly with the videos. When providers do not make captions directly available, Streasick will search online for captions where he has had success surfacing files. Catalog records are added for locally managed titles; metadata is typically sourced from OCLC, with copy cataloging done from DVD records in some cases. For short-term licenses, Streasick does add a catalog record that includes the end dates of the license term; Google Calendar alerts are used to give notice for pulling the records and files.

UM-Dearborn has a fifty title flex plan with Swank, allowing them to pay a set annual amount for access to fifty titles of their choice at any given time. Not only is this helpful from a budget planning perspective, it also allows the library to be strategic in making selections from Swank's wide portfolio of popular streaming film titles. Streasick applies a hands-on approach in managing the Swank plan, swapping titles in and out as they are needed for particular courses. By closely communicating with faculty, he will track the dates a title is needed for a course and request the title then, typically building in a week of additional time on the frontend and backend to ensure users have the access they need. Due to high fluidity of titles, they are not added to the catalog. Titles that are needed consistently are left in the plan for the full term; Streasick estimates that thirty-four films are earmarked as such, which leaves the remaining sixteen slots of the flex plan available for swapping titles at the point of need. The overall process for swapping runs smoothly for the most part; Swank is able to turn on access for requested titles within a couple of days' notice, at which point Streasick emails the film's URL to the requesting faculty member. While the process does require a lot of attention to detail for the backend management—as well as regular close communication with faculty—the overall faculty response has been understanding and positive.

Usage stats are looked at for both Kanopy and Swank, mostly for informational purposes rather than decisions on whether to renew or not. Usage for Alexander Street content is not closely looked at, as it is licensed through the University of Michigan and does not come out of UM-Dearborn's budget line; the usage stats also tend to be lumped together for all Michigan schools. Since most of the content that UM-Dearborn licenses is in response to instructor requests, the overriding assumption is that it is needed and being used in courses. Some licenses may be renewed and see low or zero usage in subsequent years, at which point Streasick will reach out to a faculty member and confirm whether it is still needed.

Regarding future challenges for managing streaming video collections, Streasick points to the high costs of individual titles. Continually needing to renew one year short-term licenses is not sustainable for library budgets. Availability of content is another challenge, as faculty will often request titles that are only streaming on commercial platforms and not licensed to libraries. Related to this challenge is managing the expectations of faculty members who assume every film is readily available to stream. Faculty tend to be understanding once it's explained to them that platforms like Netflix and Disney+ do not license content institutionally, though it remains a shared area of frustration for library users and librarians alike.

UM-Dearborn Mardigian Library, Acquisition Criteria: https://library.umd.umich.edu/services/collectionDevelopment/acquisitionCriteria.php

References

Austin, Cathy and Bonnie Thornton. 2022. "Streaming Video and Academic Libraries: A Survey on Commonalities in Acquisitions." *Journal of Electronic Resources Librarianship* 34 (3): 24555. https://doi.org/10.1080/1941126X.2022.2099040.

Cooper, Danielle M., Dylan Ruediger, and Makala Skinner. 2022. "Streaming Media Licensing and Purchasing Practices at Academic Libraries: Survey Results." *Ithaka S+R*. https://doi.org/10.18665/sr.316793.

Cross, Cheryl, Christine Fischer, and Cathy Rothermel. 2014. "Streaming Film: How to Serve Our Users." *Serials Review* 40 (3): 154–7. https://doi.org/10.1080/00987913.2014.947559.

Doi, Carolyn, James Mason, and Jared Wiercinski. 2011. "Mobile Access to Audio and Video Collections in Libraries and Other Cultural Institutions." *Partnership: The Canadian Journal of Library and Information Practice and Research* 6 (1). https://doi.org/10.21083/partnership.v6i1.1246.

Eng, Sydney and Flor A. Hernandez. 2006. "Managing Streaming Video: A New Role for Technical Services." *Library Collections, Acquisitions, & Technical Services* 30 (3–4): 214–23. https://doi.org/10.1080/14649055.2006.10766129.

farrelly, deg. 2016-17. "Issues in Academic Library Streaming Video." *Journal of Digital Media Management* 5 (2): 169–81. https://www.ingentaconnect.com/contentone/hsp/jdmm/2016/00000005/00000002/art00008#.

Laskowski, Mary S. and Thomas H. Teper. 2014. "Promoting Use and Preserving Access: Navigating the Evolving Nature of Academic Media Collections." *Library Collections, Acquisitions, & Technical Services* 38 (3–4): 54–62. https://doi.org/10.1080/14649055.2015.1055999.

Library of Congress. 2022. "Recommended Formats Statement 2022–2023." https://www.loc.gov/preservation/resources/rfs/index.html.

Schroeder, Rebecca and Julie Williamsen. 2011. "Streaming Video: The Collaborative Convergence of Technical Services, Collection Development, and Information Technology in the Academic Library." *Collection Management* 36 (2): 89–106. https://doi.org/10.1080/01462679.2011.554128.

Smith, Christine F. 2022. "Exploring Streaming Media and Course Reserves: A Case Study." *Technical Services Quarterly* 39 (3): 272–88. https://doi.org/10.1080/07317131.2022.2082661.

Tanasse, Gisèle. 2021. "Implementing and Managing Streaming Media Services in Academic Libraries." ACRL/Choice, publisher. https://www.choice360.org/wp-content/uploads/2021/10/Choice_Implementing_and_Managing_Streaming_Media_Services_in_Academic_Libraries_lo_res_100721.pdf.

Towery, Stephanie, Amanda N. Price, and Karen E. Cowen. 2019. "Video Streaming Licenses: Using a Decision Tree and Workflow Chart." *Journal of Copyright in Education and Librarianship* 3 (1): 1–32. https://doi.org/10.17161/jcel.v3i1.7483.

7 Metadata, Discovery, and Access for Streaming Video

Introduction

Discovery is a fundamental principle of library technical services work. Materials budgets are maximized to acquire content and build collections, so it naturally follows that those collections should be easy for users to find and access through multiple discovery points. The underlying principles and mechanics of discovery are no different for streaming video than they are for any other format. However, there *are* particular considerations unique to the discovery of streaming video resources that libraries will need to take into account. Metadata for streaming video can come from a number of sources, from vendor-provided machine-readable cataloging (MARC) records for larger collections to copy catalog records from OCLC WorldCat. Original cataloging work may also be required, for example, in the case of locally managed streaming files from smaller distributors. This chapter will cover considerations for making streaming video content discoverable and best practices for ensuring quality metadata.

To Catalog or Not?

It seems a given that libraries would want to maximize the discovery of their licensed streaming video content. However, some evidence suggests that decisions around the discovery of streaming video are not always so straightforward. For example, in Austin and Thornton's 2022 survey of academic libraries, it was found that 75.12 percent of respondents added film records to their OPAC and 65.9 percent made content discoverable through their discovery layer. Respondents were able to choose multiple answers, so there was some overlap in cases of libraries using both an OPAC and a discovery layer (Austin and Thornton 2022, 250–51). For libraries that use both, it is generally the case that records added to the catalog are also made visible in their discovery layer. In recent years, many libraries have moved to solely using discovery layers over catalogs. Interestingly, the numbers are largely consistent with an earlier survey of academic libraries in 2013, which found that 75 percent of respondents made streaming video content discoverable in their catalog (farrelly and Hutchison 2014, 74).

Although the use of streaming video content has grown substantially in the intervening decade between surveys, the policies toward cataloging have remained largely similar.

Given that a full quarter of libraries are not adding streaming records, it begs the question: why would a library *not* catalog its streaming videos? There are a number of potential explanations. As has been covered in previous chapters, a primary use case for acquiring streaming video content is for use in course reserves. Some libraries may solely be adding direct links for streaming videos into a learning management system (LMS) course site to limit the viewing of a specific title to students enrolled in that course. Short-term licensing is another possible driver behind not adding catalog records for streaming video. In earlier chapters, we have noted the staff effort needed to track the expiration dates for short-term licensed streaming titles, including the need to remove and in some cases delete locally managed files. Maintaining discovery records for these short-term licenses is another onus upon staff bandwidth that some libraries may not have the desire or capacity to take on. Austin and Thornton found that a full 17.51 percent of libraries surveyed chose to rely on streaming titles to be discoverable through the vendor platform interface over adding records to the catalog (Austin and Thornton 2022, 250–1). Lack of systems infrastructure can also be a barrier, as libraries may be saddled with an older integrated library systems (ILS) or be without an electronic resources management system (ERMS). In addition to staff and technological resources, financial resources may also factor into not loading catalog records. Libraries that have patron-driven acquisition (PDA) plans with one or more streaming vendors may find that those budget lines get spent down quickly and may choose to limit the discoverability of eligible titles, leaving users to discover titles by searching within the platform itself. Finally, vendor-supplied records may be non-existent or of poor quality—as we will see in the next section of this chapter—leading libraries to go without cataloging.

Regardless of whether it is decided to catalog streaming videos or not, local policies should be applied with consistency across use cases. For titles that are solely intended for course reserves, it will be necessary to ensure that accurate stable links that allow for both on-site and remote access are shared through the LMS used for reserves and by the faculty teaching the course. Brief bibliographic records may need to be maintained for linking to payment information in the library system. Even if the title is a short-term license that is not added to the library catalog for discovery by users, creating and suppressing a brief bibliographic record in the catalog will ensure that library staff can still locate the title and connect it to relevant payment information.

Streaming Video Metadata

Vendor-Supplied Records

For those libraries that do opt to make streaming video available through their own discovery platforms, records and metadata are provided in a number of different ways. The most common is to receive MARC records from the vendor which are then loaded into the library catalog. Vendor-supplied MARC records are typically freely available and often can be downloaded directly from the vendor's platform administration portal or sent by email. Depending on the collection, periodic updates may be made available as titles are added to and removed from the platform. Another potential source for copy catalog records is OCLC WorldCat. MARC records may also be made available through third-party knowledge base providers or through modern LSPs like Alma. Similarly, a library may choose to activate streaming titles through their knowledge base (KB) or central discovery index holdings, which would make them visible in the library's discovery layer. Emery, Stone, and McCracken warn against both fully cataloging streaming videos and activating them as KB targets as it could lead to conflicting results in the discovery layer. They note that for libraries that fully catalog streaming video collections, "it is often better to leave them off the central indexing within the discovery layer to ensure that content is accessible as possible" (Emery, Stone, and McCracken 2020, 77). Discovery layers are increasingly common in libraries and provided as an alternative to traditional OPACs, or in some libraries, supplanting the OPAC altogether. Their functionality for using facets to narrow search results by format is particularly helpful for users looking to discover streaming video content exclusively.

While acquiring discovery records for streaming video is straightforward more often than not, the actual quality of the records may require additional action depending on the policies of the library. Particularly in the case of vendor-supplied records, there may be significant variances in the quality of metadata. farrelly advises that trained cataloging staff evaluate vendor-supplied metadata "to determine its suitability for local needs, and arrange for other solutions" when the records fall short of local cataloging standards (farrelly 2014, 223–4). This is echoed by Breivold who notes that cataloging staff will have the expertise to determine what level of quality, detail, and customization works best locally (Breivold 2019). Open-source tools such as MarcEdit (https://marcedit.reeset.net/about-marcedit) can be useful in cases where batch edits may need to be made across a particular set of vendor-supplied records. In determining local policies for streaming video record quality, there are excellent resources developed by and for the cataloging community, such as the "Best Practices for Cataloging Streaming Media Using RDA and MARC21," a document created by OLAC, or Online Audiovisual Catalogers Inc., which is an ALA affiliate. Originally made available in 2008, then revised in 2018 and, most recently, in 2022, the "Best Practices" includes recommendations such as a provider-neutral approach to streaming media records and using separate records when a title is held in electronic and physical formats (Streaming Media RDA Guide Task Force 2022). Another resource, the ACRL Guidelines for Media Resources in Academic Libraries, most recently revised in 2018, counsels that media resources be cataloged in accordance with current national standards and that "access points should incorporate subjects, alternate titles, subtitles, translations of titles, series, and persons or bodies responsible"; accessibility features such as subtitle, caption, and audio description should also be included in the record (Association of College and

Research Libraries 2018). Additionally, a National Information Standards Organization (NISO) working group comprising stakeholders from libraries, publishers, and third-party service providers recently published guidelines for video and audio metadata that incorporate a variety of existing metadata standards, including MARC21, Dublin Core, and MODS. As a NISO-recommended practice, these guidelines establish consistent metadata practices for use across libraries, publishers, and vendors (National Information Standards Organization 2023).

A 2020 study by Wang and Loftis found that application of metadata standards was, in fact, inconsistent across vendor-supplied streaming records, with a sizable number of vendor records being cataloged with AACR2 standards instead of conforming to the more current RDA cataloging rules. The study authors went on to compare searching in their library's discovery layer versus searching directly on streaming video platforms, finding that "sometimes searches on the vendor platforms are more up to date" (Wang and Loftis 2020, 81). This is a common issue to be aware of in e-resource management and is also seen with aggregator e-book collections. Vendor lag in updating metadata across discovery sources can lead to false positives for content that has been pulled, as well as leaving some content potentially only discoverable by searching on the platform. These types of access challenges can be addressed by timely record updates on the vendor end. The same study by Wang and Loftis also noted the importance of enriching metadata to support searching for film titles in non-Latin scripts, such as Korean, as well as including alternative titles in MARC 245 fields, to ensure that a film like *Sunset Blvd.* is discoverable to users using the search string "sunset boulevard" (Wang and Loftis 2020, 83).

Metadata for Locally Hosted Content

For locally hosted or managed streaming video content, vendor-supplied catalog records may not always be available, particularly when working with smaller distributors and independent filmmakers. If the supplier does not provide metadata, catalog records could potentially be sourced through OCLC WorldCat. If a record for the DVD version of the title is available, a record for the streaming video can be derived from it using the guidelines cited above but retaining relevant content details such as the MARC 520 summary field and 650 subject headings, which generally remain the same across formats.

Given the constraints on staff bandwidth, original cataloging may not be practical for locally hosted streaming video, with the potential exception of videos that originate locally and are not acquired from an external source. However, even when using copy catalog records as a starting point, it is likely that more attention and editing may be needed than cataloging in other formats. This has long been the case with video recordings in general. A study examining the catalogs of Eastern Washington University and Gonzaga University found that to meet defined quality metadata standards, the average video recording required six edits, compared to two edits for print monographs (Myall and Chambers 2007, 245). Some of the high-level metadata for streaming video content that should always be present include:

- Title
- Publisher or rightsholder
- Publication date or copyright year
- Subject, keywords
- Summary
- URL
- Language

Another consideration for locally managed collections is that the streaming platform used to host the video files may itself utilize custom metadata fields, which could necessitate mapping the metadata from MARC records into the appropriate fields within the platform, or vice versa. Table 7.1 includes an example crosswalk for general metadata categories that may be present in a hosting platform with MARC field equivalents. A full template for creating streaming video catalog records—adapted from the OLAC "Best Practices" and used at the authors' institution—is included in Appendix C.

Table 7.1 Crosswalk for General Streaming Metadata to MARC

Streaming Platform Field	MARC Field
Title	245 $a $b
Publisher	264 $b
Rights Statement	506, 590
Date (copyright or publication year)	264 $c
Language	546
Description (including video length/duration)	300
Subject Headings, Keywords (or Tags)	6XX
Type	336, 337, 338
Alternative title	246, 740
URL	856 $u
Relation (for titles within a collection)	730
Director, Added Author Entry	710
Abstract, Summary	520
Custom Unique Identifier	Bib record ID

Discovery

Moving beyond the challenges of maintaining quality metadata for streaming video, a greater barrier may be that library users are not even searching within the library's discovery tools. A 2018 faculty survey by Lohmann and Frederiksen found that only 27.7 percent used the library catalog to discover streaming videos for teaching use, compared to 83.3 percent who used internet search engines such as Google (Lohmann and Frederiksen 2018, 106). An earlier faculty survey in 2014 found that overall, faculty considered word of mouth and doing their own searching on sites like YouTube to be more useful than the library catalog, with a portion of respondents noting that they were unaware of library discovery tools such as the catalog or subject research guides such as LibGuides (Otto 2014, 125). Students are similarly disinclined to use library discovery tools; a 2015 survey found that while 79 percent of students in higher education were using video either in class or to supplement their studies, only 32 percent reported using library tools to discover video (Leonard 2015). This disconnect demonstrates a greater need for outreach to users, not just for the marketing of collections but also for the promotion of the library discovery tools themselves.

Searching for locally hosted content, within library discovery layers and catalogs as well as on the hosting platform itself, can often entail special considerations. For example, a discovery layer may prioritize ranking for vendor-supplied content over records for local content. This was observed at Portland State University, where records for videos hosted on their locally managed platform did not appear on the first page of results from their instance of Primo discovery (Wang and Loftis 2020, 82–3). Fortunately, in this case, it was possible to modify the backend of the discovery system to give higher priority to the ranking of local records. When loading records for streaming video into discovery systems, it is important to verify the settings for the display of results and the operability of links out to the hosting platform. Settings on locally managed hosting platforms should be inspected as well. Some platforms may push out metadata for locally hosted videos publicly, making them discoverable via search engines such as Google. This was noted at the authors' institution, where e-resources staff received notices that institutional users were requesting access to locally hosted video content. It was determined that they were discovering public links through Google, links that lacked a proxy server prefix, thereby not allowing users to authenticate with their institutional credentials. Had they searched from the library catalog or course reserves systems, users would have been presented with proxied links that allowed them to authenticate and view the film. Again, this highlights the importance of educating patrons in the use of library discovery tools. Library users searching directly on local hosting platforms may also find that search results will potentially favor those titles that have full transcripts if the discovery settings index the transcript as part of the title's metadata. This is similar to situations with e-resources like aggregator databases that have a combination of abstracts and full-text content; in this case, the videos with transcripts could be considered "full-text" and therefore more likely to show up in search results. While transcripts are essential for accessibility, the settings on the platform could be modified to give greater relevance to title metadata over transcripts, or default to searching by the title overall metadata fields; this can prevent extraneous results when performing a known title search.

Access: Setup and Administration

More often than not, implementing access for streaming video is straightforward and, in the case of vendor-hosted content, similar to how access is provided for other electronic formats such as databases, e-journals, and e-book collections. In the simplest cases, the vendor will turn the access on, usually following payment confirmation, and the content will be accessible to library users across the authorized IP address ranges that the library has shared with the vendor. For libraries utilizing tools such as EZproxy to provide off-site access, it is important that the proxy IP be among the authorized IP addresses and that links in discovery records contain URLs with the proxy prefix. It is also critical to test the proxy configuration for the platform to ensure that video playback and other special features and functions work as expected with proxied access. Some third-party vendors may utilize more robust access measures that require users to authenticate via a single sign-on (SSO) mechanism such as Shibboleth. Initial setup for SSO access may require some coordination with the vendor and institutional IT or systems staff. Some e-resources require users to register their unique usernames and passwords to access content on the platform; this is most often the case for specialized resources in subject areas like business and finance which are mainly licensed to the private sector instead of for educational use. The authors are not aware of any major streaming video providers using this access model. A potential access use case for smaller distributors is hosting a film on a third-party platform such as Vimeo that is password protected. In cases like this, it is preferable to acquire a video file that can be locally held and mounted on a streaming platform managed by the library. If access credentials like a unique shared institutional password are to be managed by the library, it is important that the credentials only be shared with authorized users. One method of doing this is posting the credentials on a LibGuide or other library-maintained webpage placed behind authentication for institutional users only.

In general, the larger streaming video providers maintain administrative portals on their platforms. These portals provide a back end for library staff to accomplish various maintenance tasks such as adjusting settings on the platform's user interface like display and search, downloading MARC records to load into the library catalog, and retrieving usage statistics. When first establishing access with a new vendor that has an administrative portal, it is necessary to establish admin credentials for the library staff that will need access to the backend. In some cases, this may be a username and password that the vendor provides, or the admin login may be connected to a specific email address or addresses. Those email addresses may also serve as the point of contact for the vendor as well as the designated recipient of notification messages, such as if the platform will be down for a maintenance period. It is a best practice to designate a generic shared email address for admin and contact purposes, such as *eresources@yourlibrary.edu*. This ensures that multiple recipients see important messages and that there is no disruption in communication if library staffing turns over. Vendors may limit the admin to a single address or may allow multiple addresses to be granted admin credentials. Regardless, admin access credentials should be limited to only designated staff that need such access for their job duties. To ensure that only such staff are making changes within the admin portal, the credentials should be posted in an area only accessible to that staff, such as a designated intranet, wiki, or other local network.

Among the administrative tasks that a library will want to establish upfront on a newly licensed streaming video platform is branding. Branding options can vary among platforms; some will only allow for a text-based notification (e.g., "This resource is licensed by XYZ University") while others will allow for an institutional image or logo to display in the platform header space or homepage. The value of library branding on licensed resources is twofold. First, it provides a helpful visual confirmation that the user is accessing the content on their library's instance of the platform and that no additional authentication or sign-on is needed. Secondly, it creates awareness on the user end of the role the library plays in bringing them access to licensed resources, with all the commitment of financial costs and staff time that entails.

When streaming video content is first being added to a library's collection, confirming access is another best practice. Having library staff test out a few individual title URLs is a good idea for confirming access. This type of access checking is common among any licensed e-resources; a notable format difference with streaming is that while a PDF of an e-journal article can be quickly viewed for confirmation, an access test for a video would require pressing play and possibly watching a minute or so of the video to verify the playback quality, presence of audio and video, captions for accessibility, etc. One area to be aware of is that for some demand-driven acquisition collections, a purchase may be set to trigger at a certain point (e.g., after five minutes of playback), so take care to note this when access checking. If access is also provided through a proxied URL, the staff checking access should also replicate the path of off-site users by testing from outside the library IP range, to verify that proxied URLs are working properly and that the content provider's domain is correctly configured within tools such as EZproxy. Methods of checking proxied access while on-site include using a VPN client, a smartphone on an external network, or a terminal with a static IP address that is omitted from the proxy configuration.

Troubleshooting Access Problems

After the initial work of establishing access and creating discovery points, streaming video resources will need their access maintained in much the same way that e-resources in other formats do. The reporting process will likely be the same or similar: users who identify access issues will either self-report them or contact frontline library staff at a service desk or through online means such as email or virtual chat. A best practice for e-resource troubleshooting is to utilize a standard form to report access problems. The form should be simple enough so as not to be onerous for users to fill out (otherwise they may simply not bother) yet still gather enough information for staff to have a grasp of the problem and by what means the user was attempting to access the resource. Some of the information that should be gathered from the form includes:

- Name of resource
- URL of resource
- Was the user on-site or remote?
- Browser being used
- Referring source (i.e., was the user linking from a library discovery tool or outside search engine such as Google?)

Ideally, the form for reporting access problems will be prominently linked on the library website in multiple places, with the links easily accessible from discovery tools or other relevant library tools such as subject LibGuides. On the receiving end of the form, it should be ensured that submission results go to the relevant staff tasked with e-resources troubleshooting. Again, utilizing a generic shared email address with multiple recipients is a best practice. Depending on the library's infrastructure and complexity of operations, robust ticketing systems such as ServiceNow may be utilized for managing troubleshooting requests. The scale of the library's troubleshooting volume will be a guiding factor, but ideally, all e-resource access problems will be funneled into a single triage point, which can be as simple as a shared email address. From there, a standard protocol of actions can be followed.

The first step for e-resources troubleshooters is attempting to reproduce the user's access issue. In so doing, troubleshooters should attempt to access the resource in the same way, either from on-site or remotely, using the same browser if possible. If the troubleshooter cannot replicate the issue themselves, there is a greater possibility that the issue is being caused by a user-end hardware issue, which makes it more difficult for e-resources staff to be of assistance and more likely that the issue may be routed to IT staff. Some user-end difficulties unique to the streaming format may be due to the quality of the user's internet connection, as issues may arise with playback quality and buffering on slower connections. Some platforms' media players may not function properly in particular browsers or older versions of browsers, which is why it is important to gather this information from the user upfront. Some browser extensions or plug-ins, especially ad blockers, may interfere with playback functionality. In many instances, playback issues can be resolved by asking the end user to clear their browser's cache and cookies or use a completely different browser.

Assuming that the access issue can be replicated on the troubleshooter's end, there's a greater likelihood that the issue exists on the vendor side or within the vendor platform. Some frequently experienced access issues are:

- resource has not been paid for
- platform is down for maintenance
- content has been removed
- link is outdated or incorrect.

For each of these troubleshooting cases, general steps can be taken to diagnose the issue and respond to it, utilizing both reactive and proactive methods.

Payment Problems

It may come as a surprise to learn that unpaid invoices are one of the most common causes of e-resource access issues. Obviously, the library would not intentionally withhold payment of an e-resource without cause. Miscommunication is most often what is at fault. An invoice can get missed for any number of reasons, a prime example being just getting lost in the clutter of email inboxes, as libraries are typically receiving a steady amount of publisher email volume. Depending on how network security settings are handled at a library or institution, messages from vendors requesting payment may go into a junk or spam mail folder, indistinguishable from other email solicitations. Staff turnover can be another culprit, as vendor representatives may be accustomed to sending renewal invoices to a particular staff member who is no longer with the library. Other complications can arise with payment, as checks, which many libraries still use, can get lost in the mail. A vendor may change its payment address or banking information, which in turn means the library's business office has to update where checks or electronic fund transfers go. Ideally, the vendor would reach out to the library when an expected payment is not received before switching off access but this is not always the case.

When investigating missed payment as a cause for access problems, it should be confirmed that the lack of payment was not due to a cancellation. Sometimes during cancellation workflows, the removal of access points such as catalog records or database A-to-Z entries are missed, yielding false positives for users trying to access content that is no longer subscribed to by the library. Assuming that the library considers the subscription to be active, payment records should be searched in the library system, to determine when the resource was last paid for. This may surface a missed payment or a resource that was last paid for over a year ago, in which case a current invoice should be requested from the vendor. Subscription-based e-resources—of which many streaming video collections are—typically get invoiced on a periodic basis with enough regularity to allow for anticipation of renewal invoices. For example, if a collection is on a calendar-year term, with invoices being issued sixty days ahead of the term's end date, the library can expect to see invoices appear around November 1st. A best practice for libraries is to track renewals through extracts from their ILS/LSP or a carefully maintained spreadsheet and to reconcile renewal payment information periodically, for example on a monthly basis. This type of proactive renewals management can surface overdue invoice payments, allowing the library to stay ahead of potential access lapses as well as to prevent against other unpleasant budgetary surprises.

Platform Maintenance

Like other vendor-hosted e-resources, streaming video platforms often experience downtime due to maintenance or unexpected outages. Platform maintenance is generally scheduled and ideally communicated to customers well ahead of time. In many cases, the vendor will target maintenance windows for off-peak periods, such as late nights on weekends or during a holiday. Interestingly, since streaming videos are often used for both educational and personal viewing, these off-peak hours could in fact be prime viewing time! However, it is also likely that the scheduled maintenance windows will be a short period of time, often eight hours or less. If a library is aware that a popular streaming service will be experiencing scheduled downtime, they can communicate this message to users proactively, such as with a note on the library website or through library social media channels.

Less predictably, a streaming video provider may experience a platform outage due to unexpected issues with their host or server. On the library troubleshooting side, it can often be difficult to ascertain whether access issues are global or local. If it is suspected that the access problem is widespread, a library can quickly check this through channels such as listservs, where other libraries may send messages asking whether anyone else is experiencing the same issue with a particular platform. If the streaming provider uses social media, such as a company Twitter/X account, this may be the place to check for the latest updates on the cause of the outage and what they are doing to remedy it. Of course, reaching out directly to the vendor is always a prudent measure when troubleshooting access issues; in the case of widespread outages, it is more than likely they are already aware of the problem and using their best efforts to fix it as quickly as possible.

Missing or Removed Content

Many subscription-based streaming video collections function similarly to e-resources such as aggregator databases and e-book collections, with titles periodically dropping in and out of the collection. This is often the result of expiring rights to host content, which is more pronounced and common in the streaming video format, particularly with regard to major studio films. In the same way that films move in and out of popular streaming services like Netflix and Hulu on a regular basis, so too might films drop from a licensed collection like Swank or Kanopy. Some of the larger subscription services, such as Films on Demand or Academic Video Online, may have regularly scheduled intervals for removal of titles; it is a best practice to confirm this with the provider upfront and ensure that such removals are communicated in advance. Ensuring that collection records are updated on a regular basis is one way to prevent against users from attempting to connect to removed content through the library catalog. Of course, this also depends on the providers making an effort to ensure that their records are consistently up to date. Again, a best practice is to communicate with providers and set expectations for record updates upfront; a good time to establish benchmarks for metadata quality and update frequency is during the initial licensing process, but these areas can also be addressed or revisited at any time.

Outdated or Incorrect Links

Due to streaming videos often having short-term licenses, either when acquired through a PDA plan or a locally hosted DSL, a common variation of the above access problems is due to term expiration. In this instance, it is not a case of the content no longer being available on the platform or payment being missed for a subscription. As noted in previous chapters, short-term licensed streaming videos will often require a decision to renew and a manual ordering process to maintain access, which poses challenges from a collection management standpoint. Access issues for these titles can arise when links to short-term licensed titles are placed in an LMS or course reserves system, or an instructor's syllabus or course page. The instructor assigning the video may not be aware that the title needs to be renewed before the license term expires. Again, communication with the original requestor of the title is vital for determining whether the short-term license will need to be renewed for another term so that there are no gaps in access on the user end. Portland State University instituted a practice of including the expiration date within the link text for short-term Kanopy licenses (e.g., "Kanopy (film expires xx-xx-xxxx)") to engender greater transparency on the user end (Emery, Stone, and McCracken 2020). At Elon University, expiration dates for Kanopy titles are made publicly visible through their OCLC WMS discovery layer; an example can be seen in Figure 7.1 (Tennant 2023). A benefit of modern LSPs such as Alma, is the ability to set expiration dates for individual resources so that records will automatically be suppressed; once a short-term license expires, these links will not display to users of the library discovery tool.

Another variation of linking issues is faulty links being shared or distributed. In these cases, users are not originating from a link in the catalog or other library discovery tool, but a link that has been shared directly with them by an instructor or placed by the instructor in an LMS. Often, a user accessing a streaming video

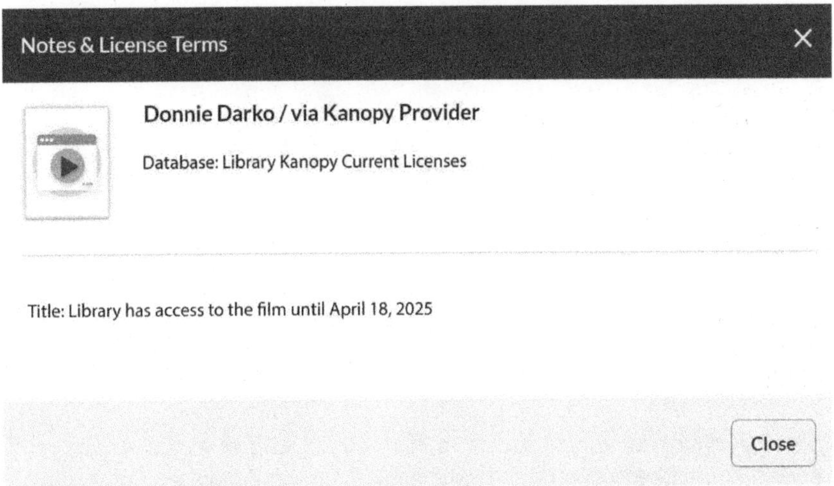

Figure 7.1 Discovery layer public note for the expiration of short-term streaming license.

will copy a link, either directly from a browser URL bar or by selecting the "copy URL" button that some platforms make available. Or they may simply bookmark that link in their browser, planning to return to it later. That link may then lack the proxy prefix in the URL which prompts remote users to authenticate. Surprisingly, this is not an uncommon problem and is frustrating to both users and instructors assigning streaming video for courses. In surveys on faculty use of streaming video, recurring difficulties were noted for links placed in the LMS (Beisler, Bucy, and Medaille 2019, 18–19) and general confusion over the need to authenticate via proxy or VPN (Horbal 2018, 181–3). Depending on the LMS configuration, it may be possible to adjust settings so that a proxy prefix is automatically appended to shared links. Similarly, the admin configuration for third-party streaming platforms may allow for links obtained through a "copy URL" function to be formatted with a proxy prefix.

Conclusion

For e-resources the processes of setting up access, ensuring quality metadata to facilitate discovery, and troubleshooting user-end problems are all interrelated. Streaming video resources share many of the common issues as other electronic formats, and libraries will similarly have to work closely with vendors to provide up-to-date discovery records for streaming collections. Access issues that are more common to the streaming format may be due to the frequency with which titles drop out of subscription collections or the tenuous nature of short-term licensed titles that will have to be renewed if they are continually needed for a particular course. Making it relatively easy for users to report access problems is essential for ensuring that these issues are addressed by library troubleshooters in a timely fashion. Overall ease of use in discovering and accessing library-provided streaming video is vital, as frustrated users may just as soon go to YouTube or other free online sources in search of video content.

Streaming Vignette—Elon University

Name of the library: Elon University (Belk Library)
Reporting librarian and title: Shannon Tennant, Coordinator of Library Collections and Associate Librarian
Type of library: Private doctoral university in Elon, North Carolina
User community: Full-time equivalent (FTE) of 7,123 students
Total streaming budget: $30,000 annually

Management of electronic resources at Elon, as well as print materials, was centralized in the Collections Department in 2019. Tennant oversees the management, including cataloging, for e-resources in all formats. The Collections budget is organized by format: e-books, e-serials, databases, web services, and other formats. Elon acquires streaming video content from many of the larger providers, including Kanopy, Docuseek, Alexander Street, and Films on Demand, which is provided through NC Live, a library consortium serving public and academic libraries in the state of North Carolina.

Budgeting for streaming video is challenging; as a smaller private institution, Elon is mainly supported by tuition dollars and does not have a large endowment. A Kanopy PDA plan was begun in 2018 and the initial budgeting was expended within six months; after that, a switch was made to a mediated plan. Currently, Kanopy purchases are mainly reserved for instructor requests, although licenses will also be purchased for campus screenings and events for titles that include PPR. Workflows specific to Kanopy are detailed and documented accordingly, with the responsibility for mediating requests delegated to an acquisitions specialist. Because Kanopy titles are heavily requested with a "high touch" workflow, there are specialized tools in place to support the request process, notably a homegrown Qualtrics form. The form collects the "needed by" date from instructors, which Tennant considers vital in order to maximize the use over the one-year license term. Recording Kanopy requests through Qualtrics also supports assessment, as fulfilled requests can be tracked and compared against actual usage, which can be cited later if low-use titles receive repeat requests. The iterative workflow for mediating Kanopy requests also builds in opportunity to see if requested content is already available through subscribed collections, thereby safeguarding against duplicate acquisitions.

When given options for short-term license lengths, the shorter length is typically chosen (e.g., a one-year term instead of a three-year term) as the more budget-conscious option. Projected use is difficult to predict over time; a professor may not teach the same course every year or they may find they want to use a different film in later iterations of the course. In some cases, Elon has been able to surface cost-savings through highly truncated short-term licenses. For example, they have licensed two-week terms for films through vendors like New Day Films. This practice is typically reserved for specialized use cases like one-time screenings for classroom viewing. While this is useful for maximizing a limited budget, it, like the Kanopy mediation, is high touch and involves staff attention and bandwidth.

In terms of collection development, the overall strategy is largely reactive, with priority given to faculty requests. The dominant model is access over ownership, with mostly subscribed collections and short-term licenses. In some rare cases, content is perpetually acquired, such as a small collection of Indian documentary films requested by a religious studies professor with a research focus in India. A set of Shakespeare videos on Bloomsbury's Drama Online platform was perpetually acquired through one-time end-of-year funds. Tennant also opts to permanently add titles from their Alexander Street Academic Video Online (AVON) collection, selecting high-use titles while filtering out short films (e.g., ten minutes or less).

Sourcing streaming video content is another challenging area. Tennant fields a number of inquiries for content on commercial platforms such as Netflix and Amazon Prime that cannot be fulfilled by the library. She notes that many students may already have access credentials and that the cost of subscribing for a couple of months is similar to or less than the costs of a textbook assigned for a course. DVDs are also acquired for course reserves and external DVD drives are loaned by the library, as most students don't have access to a DVD player. This does offer alternatives to asking students to subscribe to streaming services; still, use of the physical media options is low. Tennant also has to advise faculty members that using their own Netflix credentials to screen content in a classroom—apart from a small group of titles identified by Netflix for educational use—is not covered under fair use. JustWatch is a frequent go-to for identifying where titles can be streamed; Tennant often points faculty toward free and legal platforms, when available, although many like Tubi include advertisements. In cases where faculty pushes back and insists on an ad-less viewing experience, a licensed version will be sought out. Another challenge to sourcing is the regularity with which content drops off of platforms with little notice; faculty have to be made aware that titles, particularly those on free platforms, may not remain available for the duration of a course.

Streaming video content is made discoverable through Elon's instance of OCLC Worldshare Management System. Subscribed collections for AVON and Films on Demand are activated in the knowledgebase. Short-term licenses for Kanopy films are made available as well; dates for the license term are also added to the discovery record, so users can see when a title is nearing expiration. For the most part, however, Tennant has observed that patrons are not searching in the catalog.

Tennant foresees a continued need to utilize the streaming video budget strategically. While licensing collections like Kanopy BASE or Swank 1000 do provide access to a wide portfolio of content, faculty still prefer to request specific titles rather than choosing to assign films from a list of what is available. Continuing to license films on a title-by-title basis may remain the more cost-effective option for Elon. Films that are screened in class, as opposed to assigned for external viewing, could be acquired on DVD or through a truncated (e.g., two-week) license term. Knowing the expected use case upfront would be helpful in realizing the best potential return on investment. The budgeting practice for managing Elon's streaming video collection is high touch in terms of oversight, but it has resulted in successfully controlling costs year over year.

Belk Library, Streaming Audio & Video LibGuide: https://elon.libguides.com/streaming

References

Association of College and Research Libraries. 2018. "Guidelines for Media Resources in Academic Libraries (2018 Revision)." https://www.ala.org/acrl/standards/mediaresources.

Austin, Cathy and Bonnie Thornton. 2022. "Streaming Video and Academic Libraries: A Survey on Commonalities in Acquisitions." *Journal of Electronic Resources Librarianship* 34 (3): 245–55. https://doi.org/10.1080/1941126X.2022.2099040.

Beisler, Amalia, Rosalind Bucy, and Ann Medaille. 2019 "Streaming Video Database Features: What Do Faculty and Students Really Want?" *Journal of Electronic Resources Librarianship* 31 (1): 14–30. https://doi.org/10.1080/1941126X.2018.1562602.

Breivold, Scott. 2019. "Streaming Video Discovery in Academic Libraries." In Eric Hartnett (ed.), *Guide to Streaming Video Acquisitions*, 47–53. Chicago: ALA Editions.

Emery, Jill, Graham Stone, and Peter McCracken. 2020. *Techniques for Electronic Resource Management: TERMS and the Transition to Open*. Chicago: ALA Editions.

farrelly, deg. 2014. "Streaming Video." In Becky Albitz, Christine Avery, and Diane Zabel (eds), *Rethinking Collection Development and Management*, 215–32. Santa Barbara, CA: Libraries Unlimited.

farrelly, deg and Jane Hutchison. 2014. "ATG Special Report: Academic Library Streaming Video: Key Findings from the National Survey." *Against the Grain* 26 (5): 73–5. https://doi.org/10.7771/2380-176X.6852.

Horbal, Andy. 2018. "Instructor Use of Educational Streaming Video Resources." *The Journal of Academic Librarianship* 44 (2): 179–89. https://doi.org/10.1016/j.acalib.2018.02.009.

Leonard, Elisabeth. 2015. "Great Expectations: Students and Video in Higher Education." https://studysites.sagepub.com/repository/binaries/pdfs/StudentsandVideo.pdf.

Lohmann, Sam and Linda Frederiksen. 2018. "Faculty Awareness and Perception of Streaming Video for Teaching." *Collection Management* 43 (2): 101–19. https://doi.org/10.1080/01462679.2017.1382411.

Myall, Carolynne and Sydney Chambers. 2007. "Copy Cataloging for Print and Video Monographs in Two Academic Libraries: A Case Study of Editing Required for Accuracy and Completeness." *Cataloging & Classification Quarterly* 44 (3-4): 233–57. https://doi.org/10.1300/J104v44n03_06.

National Information Standards Organization. 2023. "Video and Audio Metadata: A Recommended Practice of the National Information Standards Organization." https://groups.niso.org/higherlogic/ws/public/download/28273/RP-41-2023_Video_Audio_Metadata_Guidelines.pdf.

Otto, Jane Johnson. 2014. "University Faculty Describe Their Use of Moving Images in Teaching and Learning and Their Perceptions of the Library's Role in That Use." *College & Research Libraries* 75 (2): 115–44. https://doi.org/10.5860/crl12-399.

Streaming Media RDA Guide Task Force. 2022. "Best Practices for Cataloging Streaming Media Using RDA and MARC21." Online Audiovisual Catalogers, Inc. https://cornerstone.lib.mnsu.edu/olac-publications/25/.

Tennant, Shannon. Interview with the authors, August 9, 2023.

Wang, Jian and Elsa Loftis. 2020. "The Library Has Infinite Streaming Content, but are Users Infinitely Content? The Library Catalog vs. Vendor Platform Discovery." *Journal of Electronic Resources Librarianship* 32 (2): 71–86. https://doi.org/10.1080/1941126X.2020.1739818.

8 Accessibility for Streaming Video

Introduction

Accessibility for library resources can be succinctly stated as ensuring that all users have equal opportunity to utilize a resource, regardless of any disability they may have. It is a core value of librarianship and permeates all aspects of collections, from the physical spaces that collections occupy to the technology used to access collections, and for the content within collections to be compatible with assistive technology. The latter aspect is key to electronic resources and will be the focus of this chapter. As libraries strive to build representative collections guided by the principles of diversity, equity, and inclusion (DEI), accessibility is another essential area of focus for ensuring that all user populations are equally served. Increasingly, libraries are recognizing and responding to this need. A 2019 survey of libraries, archives, and cultural heritage institutions found that 38 percent of those surveyed had either a formal or informal accessibility policy in place for digital content acquisition (Rosen and Grogg 2019, 13). The foundational concepts of accessibility that will be covered in this chapter have applicability to all electronic resources collections managed by libraries. For the streaming video format, the provision of captions is an integral area of focus for making content accessible. This chapter will provide an overview of best practices for making streaming video collections accessible to users.

Important Regulations and Guidelines

As a concept and workable goal, "accessibility" can be somewhat nebulous, because it is in many ways a moving target. There is almost always room for improvement in how accommodations are made, technology is utilized, and content is delivered. So while it may not be possible to make absolute distinctions between "accessible" and "not accessible" in binary terms, there are codified regulations that govern accessibility and guidelines for measuring how accessible electronic or online content is. It can be said that these regulations represent a floor rather than a ceiling—meaning a basic level of access for all users with aspirations toward further enhancing accessibility. What follows is a brief overview of some of the key regulations and guidelines that are most frequently applied to the accessibility of library resources.

For libraries based within the United States, the Americans with Disabilities Act (ADA) of 1990 is a fundamental civil rights law that prevents discrimination based on disability. Title II of the ADA specifically addresses public entities at the local and state levels, which applies to public libraries and school libraries. An important precursor to the ADA is the Rehabilitation Act of 1973, which had previously outlined accessibility requirements for institutions receiving federal funding. Per Section 504 of the Rehabilitation Act, any programs or activities receiving federal financial assistance are required to provide accommodations to individuals with disabilities. This broadly applies to educational institutions at the primary level, as well as colleges and universities both public and private, as all institutions of higher learning receive some form of federal funding. As electronic content became increasingly central to education in the years that followed the passage of these regulations, the Rehabilitation Act was amended in 1998 to reflect this. Section 508 of the Rehabilitation Act requires all electronic and information technology (EIT) that is procured with federal funding be made accessible to individuals with disabilities. For libraries in the United States that provide electronic resources to users, Section 508 is commonly invoked for the purpose of making accessibility accommodations.

Beyond US laws and regulations, a frequently utilized best practice of EIT accessibility with global scope is the Web Content Accessibility Guidelines (WCAG). Developed by the World Wide Web Consortium (W3C), the main international standards organization for the web, WCAG is a wide-ranging set of recommendations for making websites, electronic documents, and software more accessible. As of this writing, the current recommended version of the guidelines for electronic content is WCAG 2.1 Level AA. The guidelines themselves outline levels of success in accessibility areas such as keyboard shortcuts for users who cannot navigate with a mouse and color contrast for individuals who may have color blindness. WCAG's success criteria for captions and audio descriptions are of particular relevance to streaming video accessibility. The overarching principles guiding WCAG for accessibility maintain that web content is perceivable, operable, understandable, and robust (W3C Web Accessibility Initiative 2022). While WCAG was developed as an international practice, Section 508 was updated in 2018 to incorporate much of its success criteria, effectively putting it in line with WCAG 2.0 AA (LaGrow 2017).

At a more localized level, libraries in the United States may also be subject to state-specific policies that govern the use of EIT. Such policies have been put into place in states including Arizona, California, Connecticut, Illinois, Kansas, Minnesota, New York, Oklahoma, and Virginia (General Services Administration 2022). Librarians at public institutions in the United States should be aware of applicable accessibility laws and policies at the state level. At an even more granular level, there may also be institution-specific policies for making online content accessible. In an article on the roles and responsibilities of electronic resource librarians regarding accessibility, McKelvey advises awareness of relevant policies at all levels and establishing relationships with the departments and staff focused on accessibility (McKelvey 2020). At a university, this could be an office of accessibility, a disabilities service center, an assistive technology unit within an IT department, or any staff involved with remediation and providing accommodations for users with disabilities.

Another best practice for accessibility is to request that content providers supply a Voluntary Product Accessibility Template (VPAT). A VPAT is a document that indicates how third-party providers comply with Section 508 guidelines. Most of the larger content providers will be able to supply VPATs; failing the presence of a VPAT, content providers should at a minimum have some statement on their website explaining how their online platform is accessible or steps being taken toward making it accessible. It's a notable caveat that the mere presence of a VPAT is not an indicator of better accessibility or compliance with guidelines. Third-party providers are not required to complete VPATs, and the thoroughness and accuracy of VPATs are inconsistent across providers. A 2015 study by DeLancey evaluated a sample of VPATs, comparing the provider-supplied information from the VPATs to automated accessibility scans. The results found that the information in the VPATs was inaccurate 19.6 percent of the time when compared to the accessibility testing (DeLancey 2015, 106–8). While it is definitely preferable to have a VPAT over nothing, their actual utility is variable, and they may best serve as an indicator that providers are at a minimum displaying an awareness of accessibility issues.

Increasingly, non-compliance with accessibility regulations has led to legal action brought upon educational institutions. Cases filed by the US Department of Education's Office of Civil Rights against Penn State (Pennsylvania State University 2020) and the University of Montana (United States Department of Education 2014) resulted in resolutions where the institutions agreed to take steps to bring learning management systems, course materials, and library database content into compliance with Section 508. For online video content, captioning is an accessibility requirement. The National Association of the Deaf has filed lawsuits against Harvard and MIT requiring captions for content created by those institutions and made publicly available (Schaffhauser 2015). An eight-year Department of Justice investigation of UC Berkeley resulted in a consent decree wherein the institution pledged to make accessible any publicly available online content posted on their website, as well as to third-party platforms such as YouTube (United States Department of Justice 2022). In cases like these, the focus has been on video content created by the institution and made public, such as lectures recorded for a Massive Online Open Course (MOOC) or a recording of a notable speaker at a campus event. Streaming video licensed from third-party providers differs as it is not locally created, but libraries that license this content can work toward utilizing similar best practices for accessibility.

Assessing Vendor Platforms

Among libraries, there is clear recognition and value placed on the importance of accessibility. As practitioners making online content available to users, librarians often find themselves at the forefront of accessibility issues. DeLancey and Ostergaard noted the importance of the electronic resource librarian's role in accessibility, advising that "being educated about the issues, challenges, and resources available [for accessibility] is essential to serving all our patrons" (DeLancey and Ostergaard 2016, 182). E-resources librarians are naturally positioned to work on accessibility issues since they serve as a primary point of contact with vendors and distributors, work to set up access for resources, and troubleshoot user-end access issues. NASIG's *Core Competencies for Electronic Resource Librarians*, first published in 2013, is a key foundational document outlining the various expected roles of the e-resources librarian. Although accessibility is not addressed at length, the *Core Competencies* do advise familiarity with both assistive technology and regulations such as the ADA (NASIG 2013). O'Reilly has gone on to argue that accessibility is in fact a core competency for e-resource librarians, noting the impact of accessibility at various stages of the e-resource life cycle including trialing, licensing, and implementation of e-resources (O'Reilly 2020). There are numerous examples of librarians integrating accessibility into their regular assessment practices, from evaluating the accessibility of library resources (Rysavy and Michalak 2020), auditing existing library collections with a focus on third-party vendor compliance (Fernandez 2018), and incorporating accessibility practices into licensing and acquisition workflows (McKelvey 2020).

Accessibility knowledge is increasingly part of a librarian's toolkit, with a key component being the ability to make assessments of acquired content. Most licensed streaming video content is made available through third-party platforms. Ideally, these platforms will be designed to support accessibility. The following are accessibility functions for librarians to take note of when assessing a streaming platform.

Accessibility Functions—What to Look For

One of the key accessibility functions that a platform should support is captioning. Captions should present the full text of not only spoken dialogue but all audio information, including sound effects and musical cues present in the video recording. In this way, captioning differs notably from transcripts or subtitles, which only present text of dialogue. Captioning of sound effects (e.g., "a wolf howls in the distance") or music (e.g., "ominous piano notes") is valuable for conveying additional auditory information to users that provides context and may be happening off-screen or otherwise not communicated by the visuals. Captions also include speaker identification to make clear what dialogue is being spoken by whom. Captions may be open, meaning they are embedded or burned into the video file itself without the option to remove them, or closed, giving the user an option to toggle between having them off or on. For closed captions to function best, the media player controls for toggling should be clear; some users may prefer open captions on players with poorly laid-out controls. While user preferences will vary between open and closed, captioning of either kind is necessary for accessibility.

The presence of transcripts to accompany streaming video is another function that supports accessibility and usability. While often conflated with captions, transcripts can serve separate purposes for accessibility; as Spina notes, the inclusion of transcripts demonstrates how "providing multiple accessibility features also offers better, more versatile, and more flexible user experiences for everyone" (Spina 2021, 13). Transcripts may differ from captions in terms of content, as transcripts may have more robust descriptive information, such as describing what is visually being shown in the frame of the video. For this reason, transcripts are more likely to be displayed in a separate frame or sidebar of the video player, while captions are more often overlaid over the video picture itself (Kushalnagar, Lasecki, and Bigham 2013). Descriptive transcripts of this nature make video content accessible to users who are reliant on braille displays (Spina 2021, 14).

Keyboard shortcuts are another function of accessible platforms. They support users who may not be able to operate a mouse. Instead, they navigate the platform using the tab key or other pre-defined shortcuts, such as using the space bar to press play on a video. In the Kanopy media player, for example, videos can be played or paused using the space bar or enter key; the left and right arrow keys can be used to rewind or advance the video.

Audio description is an accessibility function meant to accommodate users with visual disabilities. As the name implies, it is audio describing the events occurring within the video (e.g., "a man in a suit and tie enters a crowded room"). The descriptive track plays separately from the audio within the film and is synchronized to set the scene and convey onscreen action without overlapping with the dialogue. Out of all accessibility-supporting functionality, audio description tracks are rarely made available and are not present on the majority of third-party platforms. Some distributors, such as Alexander Street and Infobase, do offer audio descriptions for select titles or may be able to provide audio descriptions at the point of need or upon request. A study by Denzer delved into which third-party vendors made audio descriptions available for their content. Denzer found that availability varied, with some vendors already taking steps to make audio-described content available. For example, using the search string "audio description" in the Academic Video Online platform will surface videos with descriptive audio (Denzer 2023). Vendor-provided catalog records for videos with descriptive audio may also display that status within the title field (e.g., "Quiet Rage: The Stanford Prison Experiment, with Audio Description").

How to Test for Accessibility

There are a number of straightforward steps that librarians can take to make some basic assessment of a streaming platform's accessibility. Chief among these is identifying the presence of captioning for video content. In doing so it is important to note that the text includes sound effects and music cues, again noting the distinction between true captions over subtitles or a transcript. Captions have a universal design benefit, regardless of whether or not a user has an auditory disability, as they allow the user to follow what is happening on screen even in cases where dialogue is hushed or spoken quickly, or there are distracting external factors making the sound difficult to hear, such as when watching video in a crowded space. A useful feature on some platforms is when they have captions or a transcript appear in a sidebar or frame separate from the picture, allowing a user to quickly scroll through the text and potentially skip ahead to a timestamp within the video containing a specific scene or piece of dialogue.

Librarians can also test how navigable a platform is without using a mouse, instead relying on the tab key to navigate across the platform, shortcut keys for play and pause functions, and if arrow keys can be used to advance or rewind the playback. Compatibility with screen reader software is another accessibility function to test for. A screen reader will read aloud the text that is present on the screen. While this assistive technology is more utilized for those with visual disabilities to access full-text e-resources such as e-books, it is still useful in video platforms to read aloud the title and credit information for films being accessed. Many libraries license screen reader programs such as JAWS; there are also freely available open-source products like NVDA. Operating systems also have built-in screen reader applications, for example, Narrator, which is built into Microsoft Windows.

The color contrast on a platform is an area that can cause accessibility issues for individuals affected by color blindness. It is important that foreground text color lightness contrasts sufficiently with the background color. Contrast and other accessibility aspects of websites can be tested with online tools such as WebAIM's WAVE web accessibility evaluation tool (https://wave.webaim.org/). The applicability here is directed toward the platform itself, as the film content may be lit in any number of ways, from excessively bright to obscurely dim. It should however be easy for users to read the external information presented on the computer screen outside the playback window, such as title and credits.

Other Resources for Assessment

Beyond these basics of accessibility checking, there are other external resources that can be consulted for evaluating a streaming video platform's accessibility. Locally, there may be an office of accessibility within the library or parent institution that has dedicated staff and resources to support users. Developing relationships with such an office can be a beneficial partnership for the library, as accessibility staff have the training to evaluate streaming platforms, assistive technology—such as screen readers—to make the resources more accessible, and additional resources to support user accommodations when necessary.

Additionally, there are larger-scale collaborations among libraries that have focused on platform accessibility. The Library Accessibility Alliance (LAA) is a partnership of libraries dedicated to equitable service and access, with a specific focus on improving electronic resource accessibility through testing, information sharing, and education.

Initially stemming from the efforts of the Big Ten Academic Alliance (BTAA)—who formed an E-Resource Accessibility Group in 2015 to begin exploring concerns of accessibility—the LAA has grown through partnerships among library consortia, beginning in 2019 when the BTAA joined efforts with the Association of Southern Research Libraries (ASERL). Since that initial partnership, the LAA has expanded its membership to include the Greater Western Library Alliance (GWLA) and the Washington Research Library Consortium (WRLC) in 2021, and in 2023 was joined by the Ivy Plus Libraries Confederation (IPLC). Some of the key formative principles that have driven the LAA's work to date have been a desire to avoid duplication of efforts among individual libraries and a realization that libraries can have a greater impact through working collectively (Pionke and Schroeder 2020, 141).

Among the resources provided by the LAA are a Library Accessibility Toolkit, a wiki-like document that provides useful terminology and current research on accessibility, model accessibility language that libraries can adopt for their e-resource licenses, and accessibility evaluations of third-party platforms (https://btaa.org/library/reports). It is this latter service that can be of particular use for libraries interested in gleaning the accessibility of their licensed e-resource content, as the growing list of evaluations covers major content providers, including popular streaming platforms such as Kanopy, Films on Demand, and Alexander Street. Evaluations are conducted by usability experts to determine how the streaming platforms adhere to WCAG 2.1 Level AA criteria. The reports identify accessibility difficulties that users will encounter, such as a lack of keyboard shortcuts, incompatibility with screen readers, and insufficient color contrast.

The magnitude of evaluating accessibility across numerous third-party platforms goes beyond that of any single library, which is what makes the work of the LAA so beneficial, particularly for libraries that may not have much in the way of dedicated accessibility resources. The LAA entries for vendor platforms are helpfully dated, so the latest reporting can be seen, and links to previous evaluations are included so that a record is maintained of how a vendor has improved their platform's accessibility. Additionally, as awareness of the LAA's efforts has grown, vendors themselves are making efforts to respond to the platform evaluations and recommendations, with this documentation also being posted to the LAA website. This is an encouraging development. Noting accessibility shortcomings in vendor platforms should not be viewed as a call out, but instead as an opportunity for libraries and vendors to collaborate and provide a more equitable user experience.

Accessibility for Locally Hosted Streaming Video

For streaming videos that are locally hosted or managed on a library-maintained platform, accessibility concerns are still very much relevant. Due to the fact that the library is itself responsible for managing the content, a proactive approach toward accessibility is necessary. The hosting platform should be assessed in the same manner that a third-party vendor's platform would be. Here again, captioning is a primary focus of these accessibility efforts and challenges can often arise when dealing with smaller distributors.

Licensing

The licensing process is the most opportune point in the acquisitions life cycle to proactively address accessibility with streaming video vendors and distributors. Most ideally, the license agreement will specifically state that the distributor provides a complete caption file in a format that is compatible with the local streaming platform in use at the library. While this is the preferred scenario for the library, locally hosted files are often acquired through smaller distributors or filmmakers, who may lack the resources and expertise to create caption files. Some distributors may only be able to provide transcript files, which, while helpful, fall short of the goal of full captions. Therefore, it is important that the license language be flexible, and also allow for the library's right to create a caption file, either made internally or through a third-party captioning service. Appendix B contains a model streaming license with examples of this accessibility verbiage. There are also a number of collaborative library groups, like the LAA, that have compiled examples of model language for accessibility that can be inserted into license agreements. Another group, the Triangle Research Libraries Network (TRLN), has created a guide for negotiating accessibility with content distributors. The guide includes example verbiage as well as advocacy strategies; for instance, it notes the importance of being able to provide specific examples of how accommodations are made, such as the creation of a caption file, which can be useful for assuaging distributors who are wary of what types of modifications are made to their licensed content (Triangle Research Libraries Network 2022).

Caption Files

When managing the caption files themselves, there are some additional considerations to take note of. The first is the file format. While there are dozens of available caption formats, two of the most common are SRT (standing for SubRip Subtitle) and VTT, or WebVTT (standing for Web Video Text Tracks). It is important to note which caption format is supported by your library's streaming platform, as one but not the other may be supported, and to request this specific file format from distributors. Some distributors may provide caption files in multiple formats, while others may favor a specific one. If a distributor is providing a caption file in an unsupported format, the library should first request that they provide the supported format. Failing that, there are freely available online tools that can be used to convert caption file formats, such as VEED and Happy Scribe. Ultimately, both SRT and VTT files are text-based files; their differences come down to file structure (e.g., SRTs require numbering for each caption, commas separate milliseconds in SRT timestamps versus periods in VTT timestamps) and functionality (VTT files are more robust, can include metadata, and allow for wider options in font size, color, and text placement).

While the VTT format allows for more functionality, SRT files are often more supported and compatible among a greater number of platforms because of their simplicity. Another common caption file format, DFXP, has formatting functionality for onscreen captions, similar to VTT, but is encoded in XML, resulting in a complex structure for the files themselves. DFXP is supported on a number of platforms, including Kaltura, Panopto, and MediaSpace, and is used by YouTube as well.

Caption files are sometimes referred to as "sidecar" files because they function alongside a parent video file. After the initial loading of the video file on a locally managed streaming platform, there will typically be an option to upload a supplemental caption or transcript file. Librarians who add caption files to their locally hosted streaming video should test them once they are added by watching a portion of the video. In addition to looking for completeness and the inclusion of sound effects and music alongside dialogue, other things to watch out for include, are the captions in sync with the video? Is the dialogue being spoken and effects cues occurring at the same time as the corresponding action on screen? Captions can fall out of sync based on the starting point of the timestamps. For example, there could be a brief onscreen lag behind the captions if the timestamp does not account for a thirty-second distributor logo animation at the start of the video. Edits could be made by adding thirty-seconds to the timestamps in the caption file or even editing the video file to remove the first thirty-seconds to bring the captions and video into alignment. However, the first course of action should be to note the discrepancy with the distributor and request accurate files.

If the distributor of a locally hosted streaming video file is unable to provide a caption file, the captions will need to be obtained from a separate source. Checking with an institutional office of accessibility, if one exists at your institution, is a good first step for captioning needs. If the captioning needs to be outsourced to a third party, a number of potential options exist. The video-hosting platform in use at the library for local management of streaming files may offer additional captioning services; this is the case for video-hosting solutions such as Panopto, which has a built-in caption request feature. Some video-hosting platforms have the functionality to automatically generate captions, although this is more often used with live streaming, and is less than optimal, as the audio transcription process often garbles proper names and acronyms, necessitating labor-intensive manual scrubbing of the captions. Human-generated captions will always be the preference. The Described and Captioned Media Program (DCMP), a non-profit organization funded by the US Department of Education and the National Association of the Deaf, has compiled a list of captioning service vendors on their website (https://dcmp.org/learn/10-captioning-service-vendors).

Unsurprisingly, manually creating captions for video content is an involved and painstaking process, and even a single caption file can be expensive, compounding the already often high costs of acquiring streaming files for local hosting. Depending on the provider, third-party captioning services can widely vary in costs, with some providers charging per minute of captioned video. Additionally, other technical considerations may impact cost, and if a faster turnaround is needed, expenses will naturally increase. While some libraries may factor accessibility expenses such as captioning into their budgets, the vast majority do not. A 2021 survey of academic libraries found that, out of 154 respondents, only 3.25 percent had the funding available to caption as needed (Tanasse 2021, 13–14). For a great many libraries, the reality will be that outsourcing of captions is done on a case-by-case basis at the point of need.

Audio Description

Audio description for streaming video entails additional complexities for content that is locally hosted. On the technical end, it is unlikely that the hosting platform will support the use of an additional audio track to play synchronously with the video file. In these cases, a new media file, one that includes the original audio as well as audio descriptions, will need to be created. As with other accessibility accommodations, this needs to be communicated up front with a distributor, and, as previously noted, the license agreement should grant the right to make such modifications to source files for the purposes of accessibility. As with captioning, audio description files can be created internally, through a local accessibility office, or outsourced to a third-party service. The DCMP also maintains a listing of description service vendors (https://dcmp.org/learn/179). Audio description of streaming video is far less common than captioning and will more than likely only be provided at the point of need for a specific accommodation.

Libraries shouldering the additional work of accessibility, is, unfortunately, indicative of the current state of affairs. Taking responsibility for captioning, whether through third-party outsourcing or manual creation internally, entails expenses both financial and in staff labor. With the vast majority of libraries lacking budgets for captioning all content, prioritization of what content is captioned may be based on identified accommodation needs and projected use of individual titles. An approach for captioning physical video used at the University of Montana utilized a tiered hierarchy, giving prioritization to first caption titles needed for individual accommodations, online courses, and course reserves (Keenan 2018). Similar approaches can be applied to streaming video, to determine high-need and high-use titles that lack captions and give them the first priority. Ultimately, the ideal is for all video content to be captioned, and part of the accessibility work of librarians is to communicate to content providers that captions, as well as transcripts and audio descriptions, are more than just added features.

Conclusion

In the past accessibility has been overlooked by libraries, but it is a vital aspect of electronic resources management. While a desire to adhere to established regulations and avoid litigation is understandable, libraries should be motivated to make their electronic content accessible as part of their mission to serve all users and uphold the principles of DEI. Accessibility should be an important consideration when acquiring streaming video content, and it can be done by evaluating the usability of platforms and verifying the presence of captions. A key aspect of this is maintaining an open dialogue with streaming vendors and distributors. In communicating the importance of accessibility, the library can both educate content providers and advocate on behalf of its users.

Streaming Vignette—University of Massachusetts Amherst

Name of the library: University of Massachusetts Amherst (W.E.B. Du Bois Library)

Reporting librarians and title: Scott Stangroom, Acquisitions Unit Coordinator; Gabriel Stetson, Coordinator of Digital Curriculum, Reserve, and Media; Benjamin Paul, Digital Reserves Administrative Specialist

Type of library: Public land-grant R1 research institution, located in Amherst, Massachusetts. Flagship University of the five-campus UMass system

User community: Full-time equivalent (FTE) of 32,229 students

Total streaming budget: Approximately $150,000 annually in continuing funds, which includes a regularly replenished mediated Kanopy deposit account

The University of Massachusetts Amherst (UMass Amherst) has an extensive collection of streaming video resources that is primarily responsive to faculty needs for curricular support. Much of the streaming content is acquired through subscription packages, although other acquisition models are also used, and in some cases, titles or collections are perpetually acquired. The streaming video collection at UMass Amherst began in earnest with dedicated funding in 2015. That same year, a Kanopy patron-driven acquisition (PDA) plan was begun that moved to a mediated model two years later when costs became unsustainable. The mediated plan operates like a deposit account—the library requests an invoice to top up the account when needed and title requests are handled by reserves staff. The library has a 150-film flex plan with Swank that allows them to swap titles in and out as needed. Subscribed content includes a number of Alexander Street collections (Academic Video Online, Dance in Video, the CBS 60 Minutes archive), Films on Demand, Naxos Video, and OntheBoards.tv, among others. Some streaming video content is acquired perpetually, including individual titles on Kanopy, locally hosted titles, and the Docuseek2 collection, which was purchased through one-time year-end discretionary funds. Because the growth of the streaming video collection is largely in response to faculty requests, there is no formal collection development policy in place. In instances where titles are perpetually licensed, the library will consider previous use—in the case of the perpetual Kanopy licenses—as well as anticipated use when weighing the decision to purchase.

Stangroom has the primary responsibility for licensing and ordering streaming video; he works closely with both Stetson and Paul in the reserves unit to coordinate requests with faculty and make links to streaming content available through the Ares reserves management system. In mediating the requests that come in for Kanopy titles, reserves staff use a number of different response templates depending on the status of the requestor. If a student is requesting a title for a specific class assignment, they will be referred back to their instructor to make the request on the student's behalf. Conversely, if a student is requesting a title for their personal research or use, reserves staff will suggest using the Boston Public Library—any Massachusetts resident may apply for a Boston Public Library e-card allowing them an alternative source for pursuing Kanopy requests. Another of the template responses from reserves directs requestors to alternate platforms; sometimes videos are duplicated in other licensed collections or available through ad-supported but otherwise free platforms like Tubi. JustWatch.com is frequently used to source platform availability for streaming titles; the VIDEOLIB listserv is also frequently helpful for posing questions of harder-to-source streaming video content. When a title is requested from a commercial platform such as Netflix, reserves staff will check to see if a DVD version is available, in some cases acquiring them from Amazon UK when the DVD is not released in the United States. If that is not a possibility, the faculty member will be given an option of having the title removed from the reserves platform or including a link to the commercial streaming site with a note that students will need to provide their own access. Stetson notes that at least one commercial streamer, Amazon Prime, has begun making their content available to license through Swank.

Accessibility is a major point of focus for streaming video collections at UMass Amherst. The library utilizes the services of 3Play Media, a company that creates captions upon request. 3Play charges for each minute of video captioned, and produces high-quality captions, guaranteeing 99 percent accuracy as well as compliance with accessibility standards such as WCAG 2.1. The library has dedicated budgeting for this service, buying minutes in bulk from 3Play that are then applied as needed for caption creation. Additionally, the library can pool efforts with the other UMass Libraries departments to share captions and transcripts. In the past, freely available caption sources like OpenSubtitles.org were investigated by UMass Amherst; however, it was found that the content from this source was often inaccurate or poorly translated, in addition to the issue with subtitles not being as robust and complete as captions. Some vendors, like Swank, have also been responsive to requests for supplying captions for titles that do not offer them by default. While efforts are made to have captions available for all video content, audio descriptions for video are made available much less frequently and only sourced when a specific request is made. For management of locally held streaming video content, UMass has a homegrown solution that is locally hosted, utilizing Able Player as the front-end media player. Able Player is highly supportive of accessibility functionality, and it allows audio description tracks to be layered with the hosted video file.

When asked about the challenges going forward with managing streaming video collections, Stetson points to costs as the primary issue. The demand for streaming content from library patrons is consistent. Even if budgets could be significantly expanded, the demand would still likewise expand to meet and surpass available funding. Compounding this challenge is the exclusivity of much of the requested content being solely available on a single platform, and in some cases, like Netflix, not available at all to license institutionally. Due to this monopolization of content, providers are able to regularly charge $500 or more for a single title, oftentimes through short-term licenses that expire after one year, requiring constant renewal and payment for continued access. This current model is unsustainable for any library, regardless of size and resources.

UMass Amherst Assistive Technology Center, W.E.B. Du Bois Library Learning Commons: https://www.umass.edu/it/assistive

References

DeLancey, Laura. 2015. "Assessing the Accuracy of Vendor-Supplied Accessibility Documentation." *Library Hi Tech* 33 (1): 103–13. https://doi.org/10.1108/LHT-08-2014-0077.

DeLancey, Laura and Kirsten Ostergaard. 2016. "Accessibility for Electronic Resources Librarians." *Serials Librarian* 71 (3–4): 180–5. https://doi.org/10.1080/0361526X.2016.1254134.

Denzer, Kelly. "A Furry Brown Bear Sits in the Snow, its Fur Speckled with White: Audio Description in the Wild." Poster presented at ER&L 2023: Electronic Resources and Libraries. Austin, Texas. March 5–9, 2023.

Fernandez, Michael. 2018. "How Accessible is Our Collection? Performing an E-Resources Accessibility Review." *Serials Librarian* 74 (1–4): 81–6. https://doi.org/10.1080/0361526X.2018.1430424.

General Services Administration. 2022 December. "State-level Accessibility Law and Policy." https://www.section508.gov/manage/laws-and-policies/state/.

Keenan, Teressa M. 2018. "Collaborating to Improve Access of Video for All." *Reference Services Review* 46 (3): 414–24. https://doi.org/10.1108/RSR-03-2018-0028.

Kushalnagar, Raja S., Walter S. Lasecki, and Jeffrey P. Bigham. "Captions Versus Transcripts for Online Video Content." Paper presented at W4A '13: International Cross-Disciplinary Conference on Web Accessibility. Rio de Janeiro, Brazil. May 13–15, 2013. https://doi.org/10.1145/2461121.2461142.

LaGrow, Martin. 2017, December 4. "The Section 508 Refresh and What It Means for Higher Education." *Educase Review*. https://er.educause.edu/articles/2017/12/the-section-508-refresh-and-what-it-means-for-higher-education.

McKelvey, Hannah. 2020. "The Role of an Electronic Resources Librarian in Procuring Accessible Online Resources." *Serials Review* 46 (2): 76–81. https://doi.org/10.1080/00987913.2020.1772012.

NASIG, Core Competencies Task Force. 2013. *NASIG Core Competencies for Electronic Resource Librarians*. https://nasig.org/Competencies-Eresources.

O'Reilly, Faye. 2020. "Digital Accessibility as a Core Competency for E-Resources Librarians." *Serials Review* 46 (2): 114–24. https://doi.org/10.1080/00987913.2020.1782631.

Pennsylvania State University. 2020. "Settlement Between Penn State University and National Federation of the Blind." https://accessibility.psu.edu/nfbpsusettlement/.

Pionke, J. J. and H. M. Schroeder. 2020. "Working Together to Improve Accessibility: Consortial E-Resource Accessibility and Advocacy." *Serials Review* 46 (2): 137–42. https://doi.org/10.1080/00987913.2020.1782630.

Rosen, Hannah and Jill Grogg. 2019 July. "Lyrasis 2019 Accessibility Survey Report." https://www.lyrasis.org/programs/Pages/Accessibility-Survey-Report.aspx.

Rysavy, Monica D. T. and Russell Michalak. 2020. "Assessing the Accessibility of Library Tools & Services When You aren't an Accessibility Expert: Part I." *Journal of Library Administration* 60 (1): 71–9. https://doi.org/10.1080/01930826.2019.1685273.

Schaffhauser, Dian. 2015, February 12. "Harvard and MIT Face Lawsuit for Lack of Online Captioning." *Campus Technology*. https://campustechnology.com/articles/2015/02/12/harvard-and-mit-face-lawsuit-for-lack-of-online-captioning.aspx.

Spina, Carli. 2021 April. "Video Accessibility." *Library Technology Reports* 57 (3). https://doi.org/10.5860/ltr.57n3.

Tanasse, Gisèle. 2021. "Implementing and Managing Streaming Media Services in Academic Libraries." ACRL/Choice, Publisher. https://www.choice360.org/wp-content/uploads/2021/10/Choice_Implementing_and_Managing_Streaming_Media_Services_in_Academic_Libraries_lo_res_100721.pdf.

Triangle Research Libraries Network. 2022. *TRLN Guide to Negotiating Accessibility in E-Resources Licenses.* bit.ly/trln-a11y-eresource-license.

United States Department of Education, Office of Civil Rights. 2014. "Resolution Agreement, University of Montana, OCR Reference No. 10122118." https://athenpro.org/sites/default/files/Univ_Montana_Missoula_FinalAgreement10_2014.pdf.

United States Department of Justice. 2022. "Consent Decree Between the United States and the Regents of the University of California, on Behalf of the University of California at Berkeley." https://www.justice.gov/opa/press-release/file/1553291/download.

W3C Web Accessibility Initiative. 2022, December 24. "Introduction to Understanding WCAG." https://www.w3.org/WAI/WCAG21/Understanding/intro.

9 Usage and Assessment of Streaming Video

Introduction

Assessment of library resources is both a periodically cyclical and constantly ongoing function of collection management. In some libraries, metrics such as usage statistics may only be looked at on an annual basis, for example, when subscription renewal decisions need to be made. Alternatively, many more libraries find themselves in a perpetual state of assessment, considering numerous high-level questions. Are collections responsive to user needs? Do resources that fall short of expected use require greater marketing and promotion? How does return on investment factor into future budget planning? Streaming video collections are no different than any other format in this regard; the same questions of assessment still apply. Given the continued growth and popularity of streaming video collections, as well as the high costs typically associated with them, assessing their usage would seemingly be a high priority for libraries. Interestingly, a 2021 survey of academic libraries found that only 7.84 percent of respondents had a formal assessment program in place for streaming media. A larger portion of those surveyed, 48.37 percent, did take play data into account for making renewal decisions (Tanasse 2021, 16–17). That still leaves a majority of libraries not assessing their streaming collections.

There are a few potential explanations for this lack of assessment when it comes to streaming video. First, streaming video is still a relatively new format for libraries, and only in recent years has there been a proliferation of vendor-offered content. In addition, many streaming video acquisitions are in response to a specific need, for example, an instructor requesting a particular film be licensed for a course and added to a learning management system (LMS) for reserves. These are typically handled as one-off purchases at the point of need. Assessing the usage of reserve requests has an aspect of redundancy, as it is already known that the title is being used for a course. Even if the title is a short-term license (e.g., one year, three years), whether or not the license is renewed will come down to whether it is again needed for a course, and not how much usage it received during the previous license term. Streaming video collections have tended to be viewed as "just in time" instead of "just in case," meaning that content is more often acquired reactively upon request, rather than proactively in anticipation of a future user need. This again is in part due to the high cost of streaming titles, when compared to physical media equivalents on DVD or Blu-ray discs. Libraries are unlikely to expend large portions of their budgets on content that may not even be used. As a result, models such as patron-driven acquisition (PDA) and evidence-based acquisition (EBA) are commonly offered for streaming video, as they guarantee that funds are going toward titles that are being used.

Increasingly, streaming vendors and distributors are packaging content in collections that are thematically organized. These can include curated collections (e.g., the Criterion Collection subset offered by Kanopy), popular films (e.g., Swank's Top 1000 collection), or specialized collections focused on documentary content (e.g., DocuSeek2). Such collections are similar to aggregator databases offered by e-resource vendors or publisher e-journal packages, and as such, naturally lend themselves to periodic evaluation and assessment when making renewal decisions. Acquisition models such as PDA also lend themselves to regular assessment, in order to make determinations of return on investment and to adjust funding levels as necessary. This chapter will present a conceptual basis for assessing streaming video collections, outline benchmarking tools, and provide both quantitative and qualitative approaches to assessment.

Quantitative Measures

The COUNTER Code of Practice

When contemplating quantitative metrics for the assessment of library resources, usage statistics are a logical starting point. By their nature, electronic resources leave a digital footprint that allows for gathering and triaging online usage. As the landscape for library content shifted from print to electronic resources, a standardized methodology for measuring e-resource usage, known as Counting Online Usage of Networked Electronic Resources (COUNTER) emerged. Begun as a collaboration between libraries, publishers, and third-party vendors in 2002, the first release of the COUNTER Code of Practice was published in 2003, and the code has continued to evolve through subsequent releases and other related initiatives.

COUNTER is now on the fifth release, or Release 5, of the Code of Practice. Release 5.1 has been recently published at the time of this writing and is set to be a requirement for content provider compliance in January 2025 (Project COUNTER 2023). The COUNTER code has continually evolved to more accurately reflect how content is delivered to users on e-resource platforms, with multiple variations of reporting that apply to specific formats such as e-journals, e-books, and databases. Subsequent developments include a protocol to automatically harvest usage reports, known as Standardized Usage Statistics Harvesting Initiative (SUSHI), removing the onus for manually downloading usage reports across multiple vendor platforms. Metrics for measuring usage of multimedia content—including audio, video, and images—were added to the fourth release, Release 4, of the COUNTER code in 2012. Two multimedia reports were made available in Release 4, the MR1 and MR2, which quantified the number of successful multimedia requests on a platform by collection, with the MR2 report including additional information for specific item types. The reporting in Release 5 changed somewhat, with an item-level master report that can be filtered by specific item types, such as journal articles and multimedia. Release 5 now offers a single standard reporting view for multimedia, the IR_M1, which displays successful multimedia requests at the item level. Many providers will offer downloadable reports through an admin portal or dashboard, seen in Figure 9.1, which allows raw usage data to be downloaded to CSV or XLS files, seen in Figure 9.2.

While COUNTER reporting has seen widespread adoption by most major publishers, it has mainly been for formats such as e-journals, e-books, and databases. The recent emergence of streaming video as a heavily used format has not corresponded with a large number of streaming providers making COUNTER-compliant usage statistics available. Of the major streaming providers, some—such as Alexander Street, Docuseek, and Films on Demand—have made COUNTER usage reports available. Other major vendors, such as Kanopy, do not provide COUNTER usage data but do present usage through customized reporting dashboards. Figure 9.3 shows the analytics data available on the Kanopy admin dashboard, which presents a top-level summary of usage over a specific time period, as well as the ability to drill down into viewing daily activity (Figure 9.4) and even view which days of the week or hours of the day experience the most user engagement (Figure 9.5). The reluctance to adopt COUNTER-compliant reporting for streaming video may in part be due to the relative newness of the format when compared to e-journals and databases. Additionally, many of the major platform's usage dashboards report data at a level of granularity that would not be easily captured in a standardized report.

Other Quantitative Metrics

Although some of the major streaming video providers lack COUNTER-compliant reporting, they instead present administrative dashboards with a range of specific usage metrics. Returning to the previous example, Kanopy provides statistics

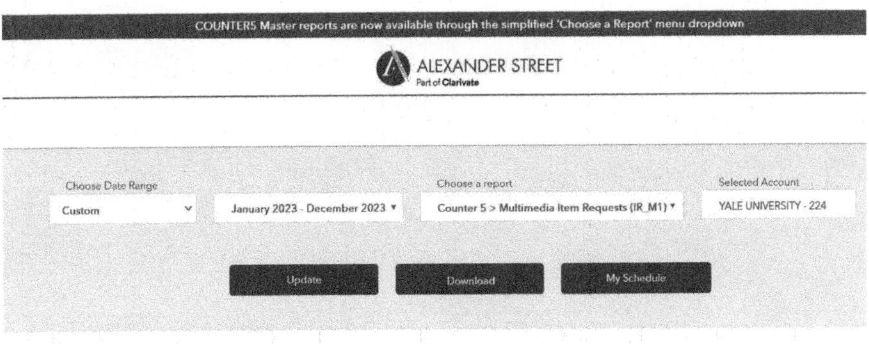

Figure 9.1 Administrative dashboard for Alexander Street platform.

Report_Name	Multimedia Item Requests		
Report_ID	IR_M1		
Release	5		
Institution_Name	YALE UNIVERSITY		
Institution_ID	ASP:224		
Metric_Types			
Report_Filters	Data_Type=Multimedia		
Report_Attributes			
Exceptions			
Reporting_Period	Begin_Date=2022-01-01;End_Date=2022-12-31		
Created	2023-08-20T17:33:15Z		
Created_By	Alexander Street Press		
Item	**Publisher**	**Publisher_ID**	**Platform**
Autism/Asperger's Conference with Dr. Temple	PESI Inc.	10415	video.alexanderstreet.com
Auto Focus	Sony Pictures Classics	12737	video.alexanderstreet.com
Autobiography of Frederick Douglass, Vol. 1	Folkways Records	5213	search.alexanderstreet.con
Autobiography of Frederick Douglass, Vol. 2	Folkways Records	5213	search.alexanderstreet.con
Aymara Leadership	Documentary Educational Resources (DER)	4124	video.alexanderstreet.com
Baadasssss!	Sony Pictures Classics	12737	video.alexanderstreet.com
Baby catchers of the Amazon	ZED (Film production)	15550	video.alexanderstreet.com
Bach English Suites	Warner Music	15032	search.alexanderstreet.con
Bach French Suites	Warner Music	15032	search.alexanderstreet.con
Bach: Cantatas, BWV 94-96	Hänssler Classic	16960	search.alexanderstreet.con
Bach: Das Wohltemperierte Clavier-Book II (CD)	Hyperion	6500	search.alexanderstreet.con
Bach: Goldberg Variations etc.	Warner Music	15032	search.alexanderstreet.con
Bach: Toccatas/Chromatic Fantasy	Musical Concepts	9364	search.alexanderstreet.con
Back to the Movement (1979âmid 80s)	Public Broadcasting Service	11145	video.alexanderstreet.com
Bad Education	Sony Pictures Classics	12737	video.alexanderstreet.com

Figure 9.2 COUNTER IR_M1 report for Alexander Street.

DOI	Proprietary_ID	URI	Metric_Type	Reporting_Period_Total	Jan-22	Feb-22	Mar-22
	video.alexanderstreet.com:3590569	3590569	Total_Item_Requests	1	0	0	1
	video.alexanderstreet.com:3360463	3360463	Total_Item_Requests	2	2	0	0
	search.alexanderstreet.com:72021	72021	Total_Item_Requests	10	0	0	0
	search.alexanderstreet.com:72025	72025	Total_Item_Requests	22	0	0	0
	video.alexanderstreet.com:764498	764498	Total_Item_Requests	2	0	0	0
	video.alexanderstreet.com:3360465	3360465	Total_Item_Requests	1	0	0	0
	video.alexanderstreet.com:1870875	1870875	Total_Item_Requests	2	0	0	0
	search.alexanderstreet.com:944152	944152	Total_Item_Requests	49	0	0	0
	search.alexanderstreet.com:967325	967325	Total_Item_Requests	7	0	0	0
	search.alexanderstreet.com:435791	435791	Total_Item_Requests	3	0	3	0
	search.alexanderstreet.com:1675675	1675675	Total_Item_Requests	7	0	0	0
	search.alexanderstreet.com:933026	933026	Total_Item_Requests	37	0	0	37
	search.alexanderstreet.com:1518596	1518596	Total_Item_Requests	3	0	0	0
	video.alexanderstreet.com:4684745	4684745	Total_Item_Requests	6	0	1	3
	video.alexanderstreet.com:5113966	5113966	Total_Item_Requests	11	0	6	0

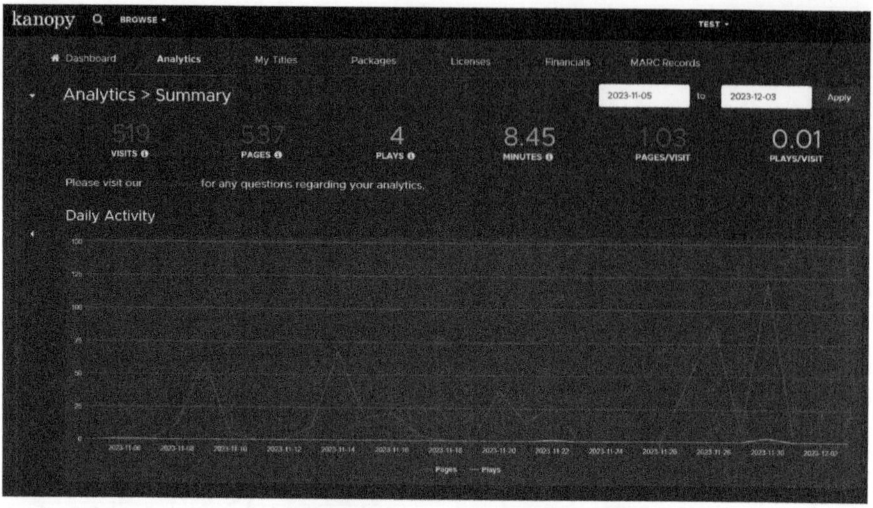

Figure 9.3 Administrative dashboard for Kanopy platform.

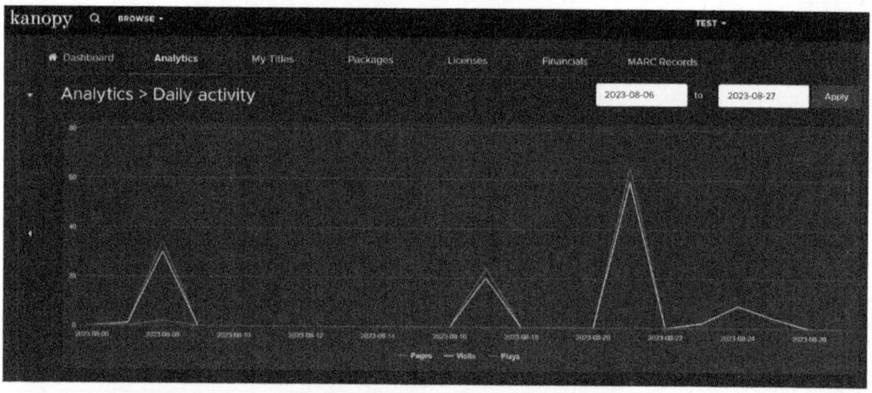

Figure 9.4 Daily activity summary on Kanopy dashboard.

Figure 9.5 Days of the week summary on Kanopy dashboard.

for a number of individual user sessions, or visits, the number of individual title pages browsed, the number of videos played, and the number of minutes viewed. Additionally, calculations are provided for the number of plays per visit or page view. These types of metrics allow for a highly granular assessment of user engagement with the platform. For individual titles, the number of minutes viewed is provided, giving insight into whether users are watching a film from start to finish, or sampling a few minutes before moving on. Viewing data is also broken out by day of the week and hour of the day, allowing for further analysis of user behavior. An uptick in views during evenings and weekends may imply that a collection is as likely to be used for leisure viewing as it is for study, a potentially noteworthy distinction for academic library collections.

Some streaming video platforms may not provide any usage statistics altogether, COUNTER or otherwise. Fortunately, sometimes quantifiable metrics can be collected from sources other than the content provider. For example, link resolver click-throughs can often be pulled from the administrative backend of a library's discovery service. If content links are maintained through LMS or LibGuides, click-through data can be gleaned from there as well. External applications like Google Analytics can potentially be configured with a library's catalog to assess referrals to specific content. While there is a limitation that these methods only provide information for users referred to the content from a specific source (e.g., a discovery layer, LibGuide), it at least provides a starting point for usage data when none is made available through the content provider.

Usage Statistics for Locally Hosted Content

For streaming videos hosted on a locally managed platform, the usage metrics made available will vary by platform. Most likely, the platform will have an administrative dashboard, similar to the bespoke dashboards for providers like Kanopy. Figure 9.6 shows the analytics dashboard for the Panopto hosting platform, which displays

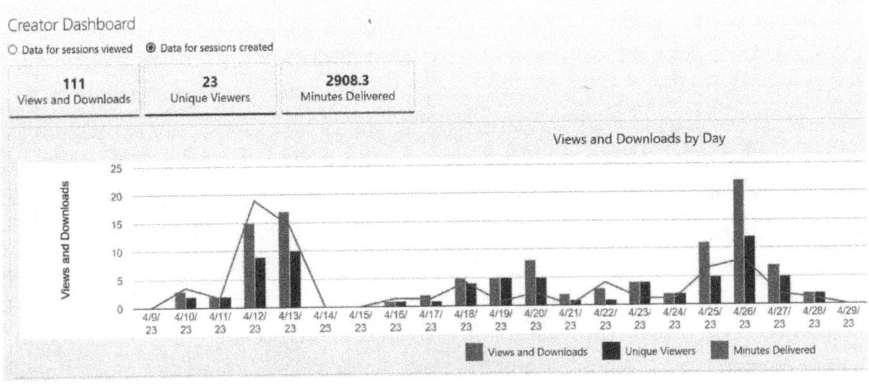

Figure 9.6 Administrative dashboard for Panopto platform.

aggregate and daily usage metrics, similar to Kanopy's dashboard. These dashboards may also provide statistics for measuring user engagement, such as unique sessions, video plays, and minutes viewed. Granular viewing information for individual titles may be available, as well as daily breakdowns of usage activity. Depending on whether the platform utilizes an authentication method such as single sign-on, unique user data may also be made available. A caveat for this data point would be that libraries should anonymize any user-specific data if they are collected, as a best practice for preserving privacy.

Cross-Platform Comparisons

In providing a standardized measurement of usage, COUNTER allows for cross-platform "apples-to-apples" comparisons across different vendor platforms. With formats such as e-books and e-journals—where adoption of COUNTER is widespread among content providers—these comparisons are straightforward. This is less the case with streaming video, where some platforms provide reports in a standardized reporting format such as COUNTER, while others utilize their own internal metrics. Considerations need to be made in order to determine corresponding metrics across platforms. Kanopy's FAQ for librarians notes that the metric "plays" should be used when making comparisons to the COUNTER metric of "item requests" (Kanopy 2023). Other platforms may have a similar metric called "views," though when making cross-platform comparisons it is important to distinguish between a video resource page being viewed and the user action of playing the video. Total minutes viewed may have utility at an individual video level, but less so in the aggregate, as the running time of videos will often vary significantly from title to title.

Calculating Assessment Measures

Cost-Per-Use

Defining what constitutes a unit of usage for videos on a streaming platform is vital to assessment, as it makes cost-per-use (CPU) calculations possible. Making such calculations is straightforward enough: simply take the cost of a resource and divide it by the number of uses in a given term to attain the CPU. It is first necessary to define what constitutes a "use" to select the proper metric for the calculation. For streaming videos, this will in all likelihood be the number of plays, or "item requests" if using COUNTER metrics. Subscribed collections with large sets of titles will likely have a reasonably low CPU when taken in the aggregate. The assessment may be different at a per title level, as e-resource collections often fall under the Pareto Principle, otherwise known as the "80/20 rule," meaning that roughly 80 percent of the collection's usage will come from 20 percent of the titles. An analysis by Nisonger looked at the applicability of the 80/20 rule across a number of library assessment measures, finding a somewhat wider dispersal among titles for electronic content usage, noting that many cases in the literature show "at least 30% of titles in electronic databases were needed to produce about 80% of usage" (Nisonger 2008, 67). Regardless of exact percentages, the overall trend is for swaths of individual titles in e-content collections to go unused, which, when visualized across a graph, appears as a trailing "long tail."

The long tail effect often presents in e-resource packages such as publisher Big Deals for e-journals and e-book collections; the development of PDA models is one way to allay concerns about large portions of content collections going unused. With PDA or any other acquisition model that is demand-driven, user need triggers the purchasing, ensuring that all purchased content is receiving usage. The CPU for PDA collections can be calculated in the aggregate and, since a standard price point is typically set for each individual video, the CPU can also be calculated at the title level. A 2014 analysis of a video PDA plan at the University of Arkansas found a wide CPU variance at the individual title level—ranging from $1.08 to $37.50 per triggered license—while the overall average CPU fell on the lower end at $5.46 (Zou and Gilbertson 2015).

CPU is a simple and straightforward measure for assessing value of collections. While CPU can provide a foundational basis for informed collection assessment, it is vital to also consider contextual factors such as the acquisition model. When perpetual access or "life of file" purchases are made—typically at an individual title level—calculating cumulative usage will demonstrate declining CPU over time and continual value of the purchase. For subscription-based packages, CPU can be informative for making renewal determinations. With the streaming video format, however, collection building is less driven by making a wide range of content available, the so-called "just-in-case" model. Instead, acquisitions tend to be targeted, often added to the collection in response to a need for a specific course. The content may receive little to no usage outside of that particular course, and as a result, have a rather low CPU. Additionally, the usage may not even be indicative of the overall viewership; for example, one video could be played during a class session for a course of thirty students and receive one use, whereas another could be assigned to view outside the class and receive thirty uses. Regardless, the overall viewership would remain the same. When assessing streaming content, it is important to treat CPU as one of many tools available and consider the context surrounding the acquisition of a particular video or collection.

Returning to the acquisition model of PDA, vendors of PDA plans will frequently tout low overall CPU figures as evidence of a good return on investment. While that may be accurate, what many libraries find is that PDA plans are also very successful at spending down allotted funds quickly. This often results in a need to re-assess the goals for the plan within the wider context of the library budget. Simply putting more money into the PDA plan is not necessarily prudent or sustainable. Many libraries have needed to suspend PDA plans (Rodgers 2019; Falloon 2020), or switch to moderated/mediated PDA plans (Foley 2018; Lowe 2020; Wang and Loftis 2020; Dotson 2022; Mason 2022) in order to exert greater control over how allotted funds are spent. This is in spite of unmediated PDA plans not triggering a purchase until a certain threshold of plays is reached, allowing the possibility of "free plays" and generally resulting in even lower CPU numbers. Mediated PDA triggers the purchase at the point of request, ensuring the initial purchase is made with that first request. Still, many libraries have opted for mediation, preferring greater control on spending to the potential for wider collection usage. A PDA plan that is a "runaway success" in terms of usage does not necessarily translate to being a good investment by the metrics of responsible collection management.

Evaluating EBA plans for success may look different, as titles are both being made available for a limited term, like a subscription, with an option for permanently licensing a subset of titles at the term's conclusion. The utility of the plan can be assessed both in terms of the collection's overall CPU and the cumulative CPU of the titles retained perpetually. For EBA models, the usage stats will inform which titles are added permanently at the conclusion of the license term, with titles receiving the highest use being the likely selections. An average cost per title figure can also be calculated by dividing the total allotted costs for the license term by the number of titles used. Table 9.1 presents a summary of assessment calculations with applicable use cases, adapted from the Spratt et al. model (Spratt et al. 2017, 187–8). The simplicity of the CPU calculation makes it flexible and highly adaptable to a number of use cases. CPU can be calculated for individually licensed titles, titles within a collection, cumulative value of perpetually licensed titles, triggered PDA titles, and collections in the aggregate. For collections, a cost per title value can also easily be arrived at by calculating the total costs and dividing by the number of titles within a collection.

Return on Investment and Cost–Benefit Analysis

Other commonly used calculations for assessing library calculations are return on investment (ROI) and cost–benefit analysis (CBA). ROI can be defined in a number of ways, and CPU figures are sometimes viewed as one type of ROI measure. A 2010 study at the University of Colorado (CU) Library system applied ROI/CBA calculations to evaluate collaborative purchasing of e-resources within the system (Pan and Fong 2010). The methods used to arrive at the calculations are instructive, with applicability for collection assessment. CBA is calculated by dividing benefits by costs (benefits/costs) and can be expressed as a ratio; anything greater than one is indicative of a net benefit over cost. When expressed as a dollar amount, the CBA number indicates the dollar amount of value or benefit gained from each $1 spent.

ROI utilizes a similar calculation but expresses the result as a percentage. The ROI formula is the difference between benefits and costs, divided by costs and then multiplied by 100, or ((Benefits-Costs)/(Costs × 100)). The ROI/CBA calculations in the 2010 study by Pan and Fong were useful in determining individual participant benefits for libraries involved in collaborative purchasing (Pan and Fong 2010, 189–91). These same formulas were utilized again at CU Libraries to compare EBA and PDA pilot programs for streaming video (Spratt et al. 2017, 188–9). The potential

Table 9.1 Assessment Measures for Streaming Video Collections

Metric	Formula	Expressed As	Use Cases
Cost-Per-Use	Cost/Use	Monetary amount	Individual term licensed titles Individual titles within collection: Cost = (Total cost of collection/Number of titles in collection); Use = (Total uses of specific title); Perpetually licensed titles: Cost/Cumulative use over time; PDA plan, individual titles: Cost = (License fee for triggered title); Collections (subscription, PDA, or EBA): Total cost/aggregate use over license term
Cost Per Title	Cost/Number of titles with usage	Monetary amount	EBA: Cost = (Total amount paid upfront) PDA: Cost = (Total amount paid for triggered titles)
Cost–Benefit Analysis	Benefits/Total Cost	Ratio (benefit amount is per $1 spent); may also be expressed as dollar amount	EBA: Benefits = (Total calculated amount of titles used); Cost = (Total or initial amount paid upfront); PDA: Benefits = (Total calculated amount of titles with usage, with or without purchase trigger); Cost = (Amount budgeted for program or initial deposit)
Return On Investment	(Benefit – Total Cost)/(Total Cost) * 100	Percentage	EBA: Benefits = (Total calculated amount of titles used); Cost = (Total or initial amount paid upfront); PDA: Benefits = (Total calculated amount of titles with usage, with or without purchase trigger); Cost = (Amount budgeted for program or initial deposit)

within PDA programs for "free plays" of videos that are viewed without a purchase trigger can result in significantly high ROI percentages. Reporting on a PDA plan at the College of Staten Island, Falloon noted yearly ROI calculations at over 2,000 percent, nonetheless reaching the cautionary conclusion that "the higher the ROI number, the more potentially [unsustainable] the program might be financially" (Falloon 2020, 51). The formulas for CBA and ROI calculations are summarized in Table 9.1, with use cases for assessing both EBA and PDA plans. The total costs or commitment for EBA plans is the initial spend at the onset of the program; for PDA, costs are the total amount of triggered licenses during the program term.

Qualitative Measures

Going beyond quantitative metrics, qualitative measures should factor into a holistic assessment of streaming video collections. A number of these measures fall under the general umbrella of vendor performance and quality of service. For example, are vendor-supplied MARC records of sufficient quality? Are updates to metadata and holdings provided in a timely manner? Is the vendor's platform subject to significant service outages where content cannot be accessed? If downtime is necessary for platform maintenance, is this communicated ahead of time, and does the period of unavailability fall within the servicing estimate? Customer service is an important qualitative measure as well. Are vendor representatives responsive to inquiries? Regarding financial matters, are invoices and credits issued within expected timelines that align with library budget planning? All of these measures can factor into whether a library may or may not decide to continue working with a particular vendor. Flexibility may be limited due to the nature of streaming video; certain content providers may have exclusive distribution rights for titles that the library wishes to acquire. However, libraries can utilize license agreements to establish a baseline of expectations with regards to performance from a vendor or distributor.

Equally deserving of consideration is the user experience with streaming video platforms. How easily can users discover and access streaming video content? For instructors, can they easily embed videos into an LMS or otherwise integrate streaming content into their teaching, by playing videos during class for example? Accessibility is another significant aspect of the user experience; methods for evaluating streaming platform accessibility were outlined in Chapter 8. Targeted surveys can also be a useful method of ascertaining how students and instructors engage with streaming video platforms, providing data points that can inform collection decisions. A 2015 survey of students at the University of Kansas found that most subscribed to one or more popular streaming services (e.g., Netflix, Hulu) and a slight majority preferred paying a fee for a digital rental over traveling to the library to use a physical course reserve. Based on the survey feedback, the library eschewed subscriptions to large collections of widely available titles (Swank and Criterion-on-Demand), instead opting to focus resources on streaming titles that were more obscure and difficult to source (Morris and Currie 2016, 494–8). Patrons who are willing to share their first-hand accounts of streaming platforms—either in a structured interview format or even just through more informal commentary—are valuable sources for assessing the user experience. The feedback received will provide insight for the library into what functionality users prefer from streaming platforms, which can in turn be communicated to content providers for future enhancements. Improving the user experience should be a shared goal for libraries and streaming video providers.

Further to that end, usability studies are another method to assess the user experience first-hand. In a 2018 usability study at the University of Nevada, Reno recruited both faculty and students to perform a set of tasks on three of the most utilized streaming platforms (Kanopy, Films on Demand, and Academic Video Online (AVON)). The testing shed light on how users engaged with each platform as well as what functionality they most preferred; faculty valued content quality and efficient searching while the students expressed a similar preference for an intuitive user interface (Beisler, Bucy, and Medaille 2019, 24–5). This echoed findings from an earlier usability study of the Kanopy and Alexander Street platforms at the University of North Texas; students and librarians preferred the user interface of Kanopy but found that functionality other than searching was complex on either platform. As a result, the library chose to pursue PDA for both platforms and focus efforts on developing their own locally managed streaming platform (Dewitt-Miller 2017).

When assessing streaming video collections, it is important to look beyond the ever-present concerns of budget and consider what purpose the library intends to achieve through the collections. Is access more important than ownership? For many libraries, the answer to this question will be "yes," as the convenience of the streaming format supersedes the concerns of opting for short-term licenses over permanently acquiring titles. This dovetails with the user's expectation for seamless and immediate access, particularly when a title is required viewing for a class. If access is the primary collection driver, then a PDA plan may be an optimal choice. Spratt et al. recommend PDA plans for libraries that prefer access to ownership and libraries with limited budgets; for libraries with greater budgets that can acquire titles perpetually, EBA is recommended (Spratt et al. 2017, 188–9). However, there is no overarching formula on what type of streaming models work best for libraries of differing sizes and budgetary means. Grombly suggests that small to medium-sized libraries instead give greater attention to focused perpetual licensing or EBA plans, arguing that the long-term cost benefits yield greater sustainability for collections (Grombly 2020, 173). Again, the library must weigh the immediate user need against stewarding a collection that will see continual usage over time. While it's a decision that all libraries must grapple with, the pull of the marketplace for streaming video is increasingly away from owned collections and toward access-based models—libraries are more often going with the tide than against it.

Discontinuing Streaming Video

In some situations, assessment will result in a library deciding to cancel, or forego acquiring streaming video content. The most straightforward example would be for streaming video content licensed via a subscription-based model. Once the term of the subscription concludes, the library has the option to renew or not renew, as they similarly would for a database or e-journal subscription. Looking at the usage statistics, the library can determine whether the ongoing cost is justifiable. Short-term licenses for individual streaming titles—whether vendor-hosted or locally hosted with a digital site license (DSL)—can also be assessed through the subscription model lens, though titles licensed specifically for course reserves may need to be renewed regardless of usage.

Additional considerations may factor into deciding whether or not to continue PDA and EBA plans for streaming video. Most EBA plans run through a pre-defined term, typically a year, at which point the access expires and the costs can be put toward permanently acquiring a number of titles based on usage. The library would determine whether or not to renew the EBA plan for another term, much in the same way they would with a subscription. This decision can be made by looking at usage across the entire collection. While CPU figures can be valuable, it is also worth assessing how many titles within the collection are seeing usage. If usage is concentrated among a small number of titles, with a long tail of unused titles, the more prudent decision may be to purchase any remaining high-use titles outright, if possible, as opposed to renewing the EBA plan for a subsequent term.

The nature of PDA plans makes a decision to continue or not more complex, as there is not a set term, but instead a set allocation as to what a library has put into a deposit account or is budgeted to spend. A PDA plan can end when the money runs out, at which point the library decides whether to deposit more funds to continue. In theory, a high enough expenditure can keep the plan running across multiple years, though the more prevalent reality, borne out by the experience of numerous libraries, is that funds will be depleted and in need of a re-up far sooner. Even if CPU figures on a PDA plan are decidedly low and indicate a good overall value, the continued costs may be untenable for the library, pushing them to move to a mediated model or to drop the plan altogether. Falloon—noting a lack of sustainability for the PDA plan at the College of Staten Island in spite of positive ROI figures—advises holistically assessing the impacts of PDA locally and concludes that "each library needs to decide their own internal benchmark number for a successful PDA program" (Falloon 2020, 51).

As is the case with any e-resource that is cancelled or discontinued, removing access to the streaming video will entail measures that should be taken leading up to and after the expiration of access. Emery, Stone, and McCracken have described this as an "exit strategy"; as the term implies, having a plan in place will engender a successful process (Emery, Stone, and McCracken 2020, 166–9). A main practical consideration is access points in the library catalog, knowledge base (KB), or other discovery products. These should be removed or deactivated so that the users do not come across resources that are no longer accessible. Some advanced communication to users may be advisable, particularly if there are titles being used by instructors for a certain course. Ideally, this will have been communicated well ahead of the access expiring; if a streaming video on request for course reserves is part of a cancellation, alternative access should be recommended, such as the use of a physical copy. As has been noted, this gets tricky to manage with individual titles licensed directly or through PDA; the library will want to ensure that a cancellation or non-renewal decision is communicated far enough in advance that the expiring title is not assigned during an interim when the access is still available but would expire before it is used in the actual course. For non-renewed titles that are locally hosted, the library will also need to take the extra steps of removal from the locally managed platform and file deletion.

Streaming video also differs from other e-resource formats in the lack of post-cancellation alternatives. If a library cancels a package of e-journals or e-books, they may still be able to procure access to individual articles or chapters through ILL. Additionally, when cancelling e-journal subscriptions, libraries often retain post-cancellation access for the subscribed years. Neither of these options is typically available for streaming video. Lending of streaming video is not being done, at the time of this writing, on a widespread level. Unlike e-journals, which are continuing resources with individual volumes and issues, a streaming video is a complete work in and of itself. Unless a title is licensed perpetually or for the life of the file, the access will end when the subscription term expires. The prevalence of access models over ownership means that streaming video acquisition is often an all-or-nothing proposition. Once the access is gone, it stays gone—or until the library decides to re-acquire—meaning that future access considerations will weigh heavily on any cancellation decisions.

Conclusion

Measures of user engagement with streaming video can take numerous forms, both quantitative and qualitative, standardized and non-standardized. While quantifiable metrics are useful for calculating the return on a library's investment, analysis of the user experience and the quality of service offered by streaming providers also contribute to a holistic assessment of streaming collections. Even collections with demonstrably high usage may not be financially sustainable for libraries, as has repeatedly been demonstrated with PDA models. In light of budgetary bottom lines, libraries may balk at licensing expensive streaming video content. Consider an individual title costing upwards of $500 or more; that film may only be used for a single course taught once a year. From the perspective of cost-per-use, it can be hard to justify the expense. However, intangibles such as user experience and the mission of the library should be considered. If a more affordably priced DVD were put on reserve, it would inconvenience users. Even buying multiple DVD copies would not assure the same sort of convenience that on demand streaming video can.

The demonstration of value or return on investment—described by Sidorko as the "Holy Grail" for libraries—can utilize multiple methodologies but will ultimately need to fit within the context of an individual institution (Sidorko 2010, 650–2). A wide array of factors is to be considered by libraries when assessing the use of streaming video collections and how best to allocate their local collection budgets. While the user demand for streaming video is likely to remain high, the current marketplace favors short-term access models, complicating collection building for the future.

Streaming Vignette—Virtual Library of Virginia

Name of the library: Virtual Library of Virginia (VIVA)
Reporting librarians and titles: Genya O'Gara, Director; Summer Durrant, Electronic Resources Librarian
Type of library: Statewide consortium
User community: Works on behalf of 71 member libraries, including public doctoral institutions, four-year colleges and universities, community and two-year branch colleges, private institutions, and other educational and research institutions including the Library of Virginia. The cumulative full-time equivalent (FTE) of institutions represented by VIVA is approximately 432,000.
Total streaming budget: Approximately 5 percent of overall collections budget

The Virtual Library of Virginia (VIVA) is a consortium that works on behalf of non-profit libraries in the state of Virginia. Collection building is the primary focus at VIVA, which differs from other state library consortia that work on shared systems infrastructure and maintenance. VIVA has a small central staff—based out of George Mason University in Fairfax, Virginia—who work in close collaboration with member libraries across the state. VIVA's annual budget in FY24 was nearly $26 million, over half of which came from member libraries, the remainder from the Virginia General Assembly. Guided by the core value of levelling the academic playing field, VIVA's shared acquisition mission ensures equitable access to licensed collections for all member libraries, regardless of size or type. This is particularly beneficial for the smaller institutions and community colleges, which do not have the budget or staffing of a large doctoral institution like the University of Virginia but can still have access to a shared set of resources.

In terms of the collection development process, VIVA has a Collections Committee with representation across all member library types. This committee meets on a quarterly basis to evaluate new offers from content providers, recommend new resources for further investigation, and evaluate existing subscriptions to determine whether they should be renewed. Recommendations go to the VIVA Steering Committee, comprising library directors, where they are either accepted or modified accordingly.

Streaming media has been a growth area in VIVA collections as of late; O'Gara noted that the budget ask consistently includes more streaming funding requests, as many member libraries have a high number of distance learners. Recent years have seen an uptick in continuing funds for streaming, in part due to the pandemic-induced shift toward distance education. VIVA's shared streaming collections include content from a wide array of content providers such as Films on Demand, Docuseek2, Alexander Street's AVON and American History in Video, Psychotherapy.net, Platino Educa, and HistoryMakers Digital Archive. An important aspect of VIVA's collection strategy is bibliodiversity, leading them to seek out partnerships with smaller providers like Platino Educa. Acquisition models for the streaming collections are generally straightforward, with most of the content being licensed on a subscription basis. AVON does have an EBA component, where some titles can be perpetually acquired at the end of a subscription term. For this case, VIVA will allocate the total perpetual access credits among member libraries, who can then select their own perpetual titles or have the credits apply to the titles with the highest use. Platino Educa also has a unique model which is pay-per-view based. In this case, VIVA sets a minimum usage cap for budgeting, then reallocates at the end of the year if usage exceeds the capping. PDA is not currently done for streaming, as it was challenging to sustainably operate a plan on a consortial level. There is a VIVA-wide contract with Kanopy that member libraries can use to run their own PDA plans.

For licensing, Durrant is involved in all negotiations and collaborates with the Collections Committee as well as individual representatives from member libraries who are delegated contacts for specific vendors. One of the major advantages of a consortium like VIVA is the ability to leverage the size of its member base to realize more favorable terms and pricing in negotiations. It is also a net positive from the vendor perspective as it reduces their workload; they can send a single invoice and the VIVA central office will rebill members as needed and handle access management, proxying, troubleshooting, etc. For access, a majority of member libraries use Alma as their LSP; VIVA centrally communicates which KB targets to activate for shared collections and also distributes MARC record files for collections as needed. In some cases, VIVA collections have customized content, and so central staff will work with third-party knowledgebase providers, such as Alma and EBSCO Discovery Service, to create targets for custom collections. These are labor-intensive to maintain and as such not preferred; VIVA makes a point of starting negotiations by looking at standard KB collections.

Assessment of collections is an area that VIVA looks at consistently and thoroughly. An internal value metric tool is used for evaluation, with both quantitative and qualitative measures being assessed. Over twenty-five individual metrics are considered, as well as additional format-specific metrics. For streaming video, it's important to compare usage of streaming collections against other streaming collections, as opposed to another format such as e-books. Additional metrics for streaming include integrative functionality with CMS and the presence of accessibility features such as audio descriptions. In using their own internal value metrics, VIVA seeks to go beyond a simplified CPU calculation to evaluate the impact and benefit of collections to their members. Returning to the value of bibliodiversity, the uniqueness of content is given high consideration. Another core value, accessibility, goes hand in hand with assessment. VIVA uses a two-tiered approach in their value metric for accessibility, both looking at the accessibility functionality of the platform itself, as well as format-specific aspects of the content. In assessing platforms, the reporting from the Library Accessibility Alliance has been immensely helpful, with most major provider platforms covered. For the format-specific metrics, much of the assessment is done in-house, such as checking for the presence of transcripts, captions, etc. O'Gara noted that streaming video providers still have a long way to go with accessibility, particularly in areas such as audio description.

In addition to accessibility, O'Gara and Durrant cited a number of other challenges with streaming video collections. Duplication of content across streaming platforms is a major challenge, particularly when keeping track of licenses, as unique content is sought after. Sustainability for streaming collections is another issue, as the current licensing models are not favorable to libraries, and rights management in areas such as public performance rights (PPR) can be murky. Finally, the overall stability of the streaming video marketplace is an ongoing problem. VIVA would like to support smaller and independent content providers, though this is challenging at a consortial level, and smaller distributors are increasingly being acquired by larger providers.

VIVA Value Metric – https://vivalib.org/va/collections/vmtf

References

Beisler, Amalia, Rosalind Bucy, and Ann Medaille. 2019. "Streaming Video Database Features: What do Faculty and Students Really Want?" *Journal of Electronic Resources Librarianship* 31 (1): 14–30. https://doi.org/10.1080/1941126X.2018.1562602.

Dewitt-Miller, Erin, Susan Smith, Xin Wang, Allyson Rodriguez, and Emily Billings. 2017. "How do I know if it's useful if I can't even get it to Open? Assessing Information Interaction at the University of North Texas to Improve Library Collections and Services." Presented at the Electronic Resources & Libraries Conference, Austin, TX, April 4. https://digital.library.unt.edu/ark:/67531/metadc977349/.

Dotson, Daniel S. 2022. "Streaming Videos: Subject Area Usage Trends." *Collection Management* 48 (2): 97–113. https://doi.org/10.1080/01462679.2022.2081831.

Emery, Jill, Graham Stone, and Peter McCracken. 2020. *Techniques for Electronic Resource Management: TERMS and the Transition to Open.* Chicago: ALA Editions.

Falloon, Kerry A. 2020. "Evaluating PDA using Financial Analysis for Streaming Videos: A Case Study at a US Academic Library." *Collection and Curation* 39 (2): 47–52. https://doi.org/10.1108/CC-03-2019–0008.

Foley, Marianne. 2018. "Streaming Video PDA: Brace Yourself, Usage is Coming." *Proceedings of the Charleston Library Conference.* http://dx.doi.org/https://doi.org/10.5703/1288284317017.

Grombly, Amanda A. 2020. "Streaming Video Acquisition Models for Small to Medium Academic Libraries." *The Serials Librarian* 79 (1–2): 163–6. https://doi.org/10.1080/0361526X.2020.1772169.

Kanopy. "Collecting Usage Data for COUNTER Reports." Last modified September 11, 2023. https://libraryhelp.kanopy.com/hc/en-us/articles/360025853934-Collecting-usage-data-for-COUNTER-reports.

Lowe, Randall A., Rick Davis, Leigh Ann DePope, Deborah K. Li, Robert K. Reeves, and Cindy Steinhoff. 2020. "Managing Streaming Video: The Experiences of Six Academic Libraries." *Journal of Electronic Resources Librarianship* 32 (2): 119–26. https://doi.org/10.1080/1941126X.2020.1739831.

Mason, Katherine F. 2022. "Managing Streaming Video License Renewals." *Serials Review* 48 (3–4): 206–7. https://doi.org/10.1080/00987913.2022.2097849.

Morris, Sara E. and Lea H. Currie. 2016. "To Stream or Not to Stream?" *New Library World* 117 (7–8): 485–98. https://doi.org/10.1108/NLW-03-2016–0021.

Nisonger, Thomas E. 2008. "The '80/20 Rule' and Core Journals." *The Serials Librarian* 55 (1–2): 62–84. https://doi.org/10.1080/03615260801970774.

Pan, Denise and Yem Fong. 2010. "Return on Investment for Collaborative Collection Development: A Cost-Benefit Evaluation of Consortia Purchasing." *Collaborative Librarianship* 2 (4): 183–92. https://doi.org/10.29087/2010.2.4.06.

Project COUNTER. 2023. "COUNTER Code of Practice Release 5.1." https://cop5.projectcounter.org/en/5.1/.

Rodgers, Andrea. 2019. "Once Upon a Time in Streaming Video: A Community College's Adventure with Kanopy's PDA Model." *College & Research Library News* 80 (9): 498–501, 505. https://doi.org/10.5860/crln.80.8.498.

Sidorko, Peter Edward. 2010. "Demonstrating ROI in the Library: The Holy Grail Search Continues." *Library Management* 31 (8–9): 645–53. https://doi.org/10.1108/01435121011093414.

Spratt, Stephanie J., Gabrielle Wiersma, Rhonda Glazier, and Denise Pan. 2017. "Exploring the Evidence in Evidence-Based Acquisition." *The Serials Librarian* 72 (1–4): 183–9. https://doi.org/10.1080/0361526X.2017.1321901.

Tanasse, Gisèle. 2021. "Implementing and Managing Streaming Media Services in Academic Libraries." ACRL/Choice paper. https://www.choice360.org/wp-content/uploads/2021/10/Choice_Implementing_and_Managing_Streaming_Media_Services_in_Academic_Libraries_lo_res_100721.pdf.

Wang, Jian and Elsa Loftis. 2020. "The Library has Infinite Streaming Content, but are Users Infinitely Content? The Library Catalog vs. Vendor Platform Discovery." *Journal of Electronic Resources Librarianship* 32 (2): 71–86. https://doi.org/10.1080/1941126X.2020.1739818.

Zou, Tim and Mary Gilbertson. 2015. "Demystifying the DDA Acquisition Model for Online Videos: In the University of Arkansas Case." Presented at the National Media Market Annual Meeting, Albuquerque, New Mexico, October.

10 Future Directions for Streaming Video in Libraries

Introduction

While making any sort of long-term prognosis on the future of streaming video is inherently foolish, the authors will offer this not-so-bold prediction: it is not going away. How we consume and engage with streaming video may look very differently in just a few years, but it will still be deeply intertwined in our lives. The number of streaming video users in the United States was estimated to be 157 million in 2023 and is projected to steadily climb to 180 million by 2027 (Statista 2023). Even compared to five years ago, the increased ubiquity of online video in our lives feels like a significant change. The prevalence of using online video for work, education, entertainment, and socializing was further intensified by the COVID-19 pandemic, which demonstrated that utilizing online video was sometimes necessary, sometimes more productive than in-person, and sometimes even preferable.

"The Streaming Wars"

It can be convincingly said that, at the time of this writing, the streaming video landscape is at an inflection point. Granted, that same generalization could have been made at any point in the last decade and still have been accurate to one degree or another. What makes the current moment different is the uncertainty of a streaming marketplace that is in a state of flux. Entire books have already been written chronicling the so-called "Streaming Wars" (Hayes and Chmielewski 2022), the battle between major media corporations for streaming users and revenue. This melee arguably reached peak intensity during 2019–20, when several global media empires propelled their platforms into a quickly crowding marketplace: Disney+, Apple TV+, HBO Max (now simply "Max"), NBC Universal's Peacock, and the soon-to-be historical footnote/punchline Quibi (which burned not so brightly for six months before shutting down)—all of these platforms were launched within six months between November 2019 and April 2020 (Kenworthy 2020).

One of the initial promising aspects of the streaming explosion was cord-cutting—the savings potential of dropping costly cable packages in favor of more reasonably priced streaming subscriptions. While the rise of streaming platforms did coincide with a decline in cable television subscriptions, would-be cord-cutters found themselves quickly inundated with numerous platforms to choose from. These platforms distinguished themselves through their exclusive content offerings; a viewer who wanted to see the latest episodes of *Stranger Things*, *Succession*, and *The Mandalorian* would therefore need to subscribe to three separate services (Netflix, Max, and Disney+, respectively). The aggregate costs of multiple à la carte streaming subscriptions could easily be greater than what was being spent on cable. As a result, consumers became more selective in their streaming choices, leading many media analysts to declare the streaming wars "over" soon after they began (Pallotta 2022; Kafka and Molla 2023; Berman 2023). Faced with dwindling profit margins, media giants like HBO and Disney began cancelling series, pulling content from their platforms, and even licensing out existing intellectual properties to other platforms. Netflix too had to tighten their belts, rescinding their previous *laissez-faire* attitude on password sharing and cracking down on users, as well as introducing ad-based subscription plans. As the streaming wars reached their attrition phase, the Writers Guild of America went on a protracted strike throughout the Summer of 2023. Streaming media was among the motivating factors of the strike, as content creators sought to receive more fair residual payments from streaming, a burgeoning revenue source that has simultaneously cut into the traditional old media revenue sources of syndication and DVD sale residuals (Koblin and Barnes 2023). While there is an atmosphere of overall uncertainty, the overarching projection is that the streaming landscape will continue to consolidate, leaving only the largest companies in business; what that will look like for consumers is still anyone's guess (Nathanson 2022).

So what will this mean for libraries? Library patrons may inquire about a popular streaming title in the same way they would the latest bestseller. A Netflix series like *Bridgerton* may be requested by a public library patron for entertainment purposes or a cultural studies professor for instructional purposes. Libraries have no recourse for providing such titles due to the streaming companies' long-standing reticence to institutionally licensing content. In the current climate of needing to maximize their individual user bases, none of this is likely to change.

Outlook for the Library Streaming Market

Inside the niche of the library world, although the library streaming video marketplace does not include popular subscription-based platforms like Netflix, the presence of a small number of major providers suggests the potential for similar consolidation. In 2021, OverDrive—already the largest provider of popular e-books and audiobooks—acquired Kanopy, the fastest-growing streaming platform in the library marketplace. Cooper and Schonfeld posited that the acquisition could herald a new empire in the public library space, where OverDrive had already achieved great success as an e-book and audiobook platform, and saw the potential for OverDrive to make greater inroads in academic libraries (Cooper and Schonfeld 2021). Another major streaming provider, Alexander Street, is under the umbrella of Clarivate, which itself acquired ProQuest in 2021, and boasts a wide-ranging portfolio as both a content provider with ProQuest and a systems provider in its ownership of Ex Libris. Like the library marketplace at large, streaming video in libraries is increasingly dominated by the same major players. With e-resource formats, industry consolidation is nothing new. There are a handful of major e-book publishers, while the largest e-journal publishers regularly acquire the portfolios of smaller society and academic publishing houses. Looking past the largest streaming video providers, there are only a few providers that could be termed mid-level or medium-sized. Increasingly, these midrange providers have worked with the larger platforms; both Kino Lorber and Women Make Movies have partnered with Kanopy to make collections of their titles available to stream on the Kanopy platform. Such partnerships make logical business sense; Kanopy can provide wider visibility for the portfolios of smaller providers and offer the infrastructure of a platform that will be familiar to library users. How other large academic publishers can influence the streaming video market for libraries remains unclear. With few notable exceptions, such as the Journal of Visualized Experiments and Sage Research Methods Video, academic publishers have not leveraged video to the degree they potentially could (Iglesias 2020).

While content providers devise new strategies on how to make streaming video available and build up their offerings, libraries have been grappling with how best to acquire existing content on limited budgets. Libraries recognize both the necessity and demand of the streaming video format, as borne out in multiple surveys. A 2021 survey of libraries found that approximately one-third anticipated a "slight" or "significant increase" in streaming media budgets, while another 46 percent forecasted they would remain steady (Tanasse 2021, 11–12). Another survey of library directors forecasts that by 2025, streaming media spending will increase to comprise an 8 percent share of materials budgets, up from 5 percent in 2020 (Cooper, Ruediger, and Skinner 2022, 12–13). At the same time, libraries are reactively responding to the need for greater control over their streaming budgets, as existing acquisition models like patron-driven acquisitions (PDA) quickly overspent allocations, and many libraries switched to mediating their plans (Foley 2018; Urban 2022; Dotson 2022; Mason 2022). For their part, vendors have begun to recognize the budgetary strain and expanded their selection of acquisition models to better suit library needs. The most noteworthy example is Kanopy, which now offers a fixed-price subscription package, fittingly branded Kanopy BASE (Bundled Academic Subscription for Education) as well as greater PDA plan customizability with the Smart PDA and Capped PDA models ("Kanopy Launches New and More Affordable Acquisition Models" 2020). Nonetheless, as available streaming video content continues to expand, streaming collections will continually require close stewardship, with library collection managers maintaining a high degree of vigilance over budgets.

Major Problem Areas

Of course, budgetary management as a present and future challenge is nothing new, and it applies across all formats. Throughout this book, we have identified challenges unique to the management of streaming video collections. We would like to conclude the book by spotlighting three of the most prevalent challenges for libraries: unavailability of content to license, inability to share content, and instability of content for long-term preservation. While none of these are new challenges, they are both of pressing concern to librarians who work with streaming video and also unlikely to be soon resolved in a definitive or expedient manner.

Unavailability of Streaming Video Content

As noted throughout this book and earlier in this chapter, the most prominent examples of streaming video content being unavailable to libraries are the films and series that are exclusive to subscription-based platforms such as Netflix and Hulu. With few exceptions, there has been little traction gained by libraries in this area. Netflix does allow for some of its documentary holdings to be screened in classrooms via one-time educational screenings, but this is limited to once per semester and a select group of titles. The downturn in profitability outlined above suggests that the bubble may have burst for streaming services and that hanging onto subscribers is imperative. None of this bodes well for libraries hoping to license exclusive streaming content for their users. One small but positive development in this area was the news that the vendor Swank has partnered with Amazon Studios to license their films through the Swank Digital Campus platform. This would make films previously exclusive to stream on Amazon Prime, such as *Manchester by the Sea,* available to license for academic use.

When videos are assigned for academic coursework, the costs are often passed along to the students, who may need to take on one or more streaming subscriptions in a given semester, in addition to the increasingly high cost of textbooks. Students concentrating in areas such as film studies or cultural studies would be particularly impacted, although video is increasingly used throughout the humanities and beyond. A 2018 study by Rodgers surveyed film studies students at Canadian universities and found overwhelming preferences of respondents wanting to watch films for free and on-demand, or "according to [their] own schedule" (Rodgers 2018, 575). Legality was less of a concern to the students surveyed, with many respondents opting to download films through file-sharing sites or stream over free websites, where content is not always necessarily uploaded by the rights holder (Rodgers 2018, 575–576). While the clampdown on password sharing by services like Netflix may negate the dubious loophole of users "borrowing" login credentials from family and friends, other extra-legal avenues will always be available to intrepid viewers who do not wish to pay. Of course, the vast majority of users would prefer access to streaming content to be both free and legal; the library is unfortunately shut out as a would-be facilitator since streaming providers prefer users to subscribe directly. For individual titles, one-time rentals may be an option. Similar to the era when video chains like Blockbuster were at their peak, these rentals allow users to view video on a personal device at a reasonable price point, often $5 or less, for a limited period typically between twenty-four and forty-eight hours. While this can satiate the need for casual viewing, students increasingly utilize film in the same way they would a novel or any other text, requiring repeat viewings or close readings of specific sequences. Multiple viewings may entail multiple rentals or a digital download purchase at a higher price point. Another issue is the affordability of educational materials for students, a foundational value for libraries to support, one which is out of reach when streaming video is unavailable to license.

Related to the user preference for the convenience of streaming video on demand is the user expectation that all video content is available to stream. In reality, there is a vast amount of content that is not readily available online, much less available to license by libraries. At the authors' institution, the move to replace physical media on reserves with streaming equivalents was able to surface options to institutionally license 56 percent of the reserves titles (Serrano and Fernandez 2023, 12–13). Some of the remaining 44 percent were available online but were unavailable for licensing or on a free site like Tubi where the content may be pulled anytime. The complexity of ownership and distribution rights for film can cause many films to go "missing" and not be readily available through any formats except legacy media like VHS. It is not necessarily older or obscure films either; titles like Mary Harron's *I Shot Andy Warhol*, Elaine May's original version of *The Heartbreak Kid*, and Julie Taymor's adaptation of *Titus Andronicus*, starring Anthony Hopkins, are all among the missing movies unavailable to stream ("List of Missing Movies," n.d.).

In academic libraries, there is increasing demand for specialized titles, oftentimes documentary content or content made outside of the United States. These titles can be difficult to source as well and may involve identifying a small distributor or working directly with the filmmakers themselves. When the content can be licensed, it is frequently in the form of digital source files that entail their own added work of local hosting and management, which was covered in detail in Chapter 6. Determining how a title can be streamed, or if it is even available, can lead to questions of ownership and distribution rights, which may be murky, at best.

Inability to Share Streaming Video Content

Assuming that a particular streaming video can be licensed for library use, a not unreasonable question to ask is, can it be loaned to other libraries? After all, collective sharing of knowledge is a principle central to the mission of libraries, and scholarly sharing of e-journal articles and e-books (at least at the chapter level) is a common practice through interlibrary loan (ILL). For streaming video, however, this has not been the case, with almost no exceptions. The reasons for this largely stem from matters of legality and technology.

On the legal and copyright front, streaming video licenses do not contain ILL clauses in the way that an e-journal license may. This again enters the realm of ownership and distribution rights; in many or most cases, the streaming vendor licensing the content is not the rights holder and is granting the permission to stream content for a limited term, directly to an individual library. Naturally, profit motivation is another driving factor for streaming vendors to not allow lending to other institutions. If one library is able to lend streaming content to another library, it undercuts potential profits for the vendors who could license the content to both libraries. While ILL can be on the negotiation table like any other term of use during the licensing process, there is little evidence that libraries would push for it or consider it a deal breaker when a streaming vendor decrees that it cannot be accommodated. Another major reason why lending rights are not a part of streaming video licenses: there is no real established practice for lending streaming video.

This lack of established practice dovetails with the technological challenges of lending streaming video. The hurdles to overcome encompass questions of file format and size, viewing platform, as well as user identity management and access control. A single e-journal article or e-book chapter can easily be delivered to a user in highly portable file formats like PDF or EPUB. An MP4 file of a digital video would be more difficult to transmit due to file size. Moreover, it would also be a non-starter for content distributors wary of digital piracy, given the past twenty-five years of post-Napster Era efforts to lock down the peer-to-peer sharing of valuable intellectual property that can be easily copied and distributed as digital files. Any progress to be made in the space of ILL for streaming video would require a high level of coordination between libraries and content providers to exert sufficiently secure access controls.

Interestingly, these challenges mirror those of libraries trying to pursue whole e-book lending. A framework known as Controlled Digital Lending (CDL) has recently gained traction as a means to address both the copyright and technical issues related to loaning digital copies of e-books. On the legal front, CDL utilizes a good faith application of two principal components of US copyright law: fair use and the first sale doctrine. With regards to the technical implementation, "properly implemented CDL programs maintain an 'owned to loaned' ratio that it is comparable to physical lending" (Bailey et al. 2018). So if a library owns two copies of a book, at any given time one physical copy and one digital "copy" can be on loan; or if two digital copies are on loan, both physical copies cannot be loaned. In a White Paper by Hansen and Courtney, the authors argue that CDL negates potential market harm to publishers through maintaining the "owned-to-loaned" ratio; it is not a radical departure in lending, CDL simply "[shifts] lending to a new format that opens up access possibilities" (Hansen and Courtney 2018).

Could the principles of CDL for e-books have applications for streaming video? For copyright law, the fair use and first sale doctrine can apply to video content, and a library owning films in physical formats could still maintain an "owned-to-loaned" ratio with digital copies. The greatest stumbling block, at least for libraries in the United States, may be getting the digital copy in the first place, as this hits against Title 17 of the US copyright code, the Digital Millennium Copyright Act (DMCA). The DMCA forbids circumventing encryption measures that are present on most DVDs, essentially preventing the ripping of a digital file for streaming. Lear characterizes DMCA as "a difficult challenge in implementing CDL for video content" and posits short- and long-term solutions (Lear 2022, 15). In the short-term, screen capture software could be used as an alternative to file ripping software that would not circumvent DMCA—although this may lead to degradation of the image quality. The longer-term solution—and no doubt a more Quixotic one in attempting—is to advocate to the US Copyright Office and by extension Congress to make permanent exceptions to the DMCA (Lear 2022, 15–16). A factor of the fair use principle that CDL rests upon is the use for educational purposes, which is not necessarily clear-cut where streaming video is concerned. Popular films like *Barbie* or *Oppenheimer* could well likely be assigned for academic coursework but could also be consumed as leisure viewing, a point of distinction that may not arise for a documentary on J. Robert Oppenheimer or the manufacturing and cultural history of Barbie dolls (2018's *Tiny Shoulders: Rethinking Barbie* is a Hulu exclusive, alas). Ensuring CDL for video constitutes educational use would likely require a high degree of administrative management "in order to provide an additional layer of protection that only allows those users who require the content for a class" (Lear 2022, 16).

While CDL offers a potential, if still complex, pathway to lending streaming video, it is nonetheless dependent on ownership of a physical copy. Most streaming video content is licensed solely through online platforms with few of those titles being held in physical equivalents by libraries. Lending of streaming video content on the most prominent platforms has yet to be attempted on a wide scale. On a statewide consortium level, notable advances have been made by the Colorado Alliance of Research Libraries, known as "The Alliance," whose membership comprises fifteen academic institutions and the Denver Public Library (Colorado Alliance of Research Libraries n.d.). At the start of 2020, The Alliance began a pilot of a project called Streaming Interlibrary Loan Video Resources (SILLVR), partnering with the streaming distributors—Swank and Films on Demand—to make video content discoverable and lendable to Alliance patrons across the consortium's shared catalog and ILL portal (DiVittorio et al. 2020). While Films on Demand put their participation on hold after a year, the second year of the SILLVR pilot saw the continued participation of Swank and the addition of Alexander Street as a partnering vendor. Speaking at a conference in 2021, Rose Nelson, the Assistant Director for The Alliance, stressed the importance of being persistent in requesting ILL permissions from vendors, noting that increases in subscription revenue for participating vendors can be an incentive for new vendor partners to join (DiVittorio et al 2022). Projects like SILLVR are an encouraging development and provide a model for consortial partnerships to expand access to streaming video collections. Institutional lending both supports a core value of libraries and also helps mitigate the budgetary challenges of the streaming video format. In the first year of the SILLVR pilot, 95 percent of requests came from public libraries, greatly expanding the reach of streaming video to libraries that otherwise may not have been able to afford it (DiVittorio et al. 2022, 37).

Seeing some major streaming vendors inching more toward permitting the lending of videos is undeniably a positive development; however, there is still a wide swath of more niche and specialized titles acquired by libraries that often end up hosted as locally managed files. The demand for lending these titles may in a way be higher or more pressing, as they are held by a limited number of libraries. Typically, they are licensed through smaller distributors or directly with the original filmmaker or rights holder. The potential benefit of lending is that the rights holder may be more amenable to a license that includes an ILL clause. Interestingly, locally managed content may also be more conducive for lending, as the institution can exert greater administrative control over the hosting platform and the security of access permissions. At the author's institution, the CMS platform Aviary is in use for video hosting, which allows a high amount of granularity in access control, such as the creation of temporary user credentials that can be assigned at a collection or individual item level. In that way, an outside user can be granted access to a specific video, with the access set to automatically expire after five days or any other pre-determined loan period. As libraries continue to grow specialized collections of locally held streaming video content, this combined focus on licensing lending rights and managing access permissions can expose the visibility of and broaden access to specialized video collections. In doing so, however, a respectful consideration of the rights holder should be a moderating factor, as these titles are made available by smaller distributors who rely more on individual sales than a well-established vendor like Alexander Street would.

Instability of Streaming Video as a Format

As caretakers of the cultural record, libraries have a responsibility for preserving collections. Streaming video will no doubt comprise a major share of library collections in the future. However, the fragility of streaming video as a format makes its long-term or future prospects less certain. One previously noted issue is simple unavailability; the common assumption that any film is available to stream is fallacious. A film may be unavailable on streaming platforms for a number of reasons. The current distribution rights may be in legal limbo. Even when there is a clear rights holder, that distributor may opt not to convert or host titles for streaming that they deem unprofitable. The non-profit organization, Missing Movies, was formed in 2022 in response to the issue of orphaned film content, with a focus on both the complex legal issues of researching and determining contractual rights that may be long expired or held by companies that no longer exist, as well as the technical work of sourcing film masters and creating new high-definition digital transfers ("The 'Missing Movies' Manifesto" n.d.). One of the co-founders of Missing Movies, Amy Heller, describes navigating the rights landscape thusly: "It takes a village of lawyers, copyright experts, archivists, distributors, and sometimes bankruptcy court judges to disentangle these very complicated messes" (Rottenberg 2023). Through such laborious efforts, Missing Movies has already "found" and cleared the streaming rights for acclaimed titles such as *Kiss of the Spider Woman* and *Mississippi Masala*.

Even titles that have a home on streaming platforms can be in a precarious position. Existing content can be pulled off a platform overnight, an increasingly common occurrence as content providers seek to cut expenses (Andrews 2022). Shifts in cultural norms can lead to individual episodes of streaming series with problematic subject matter getting pulled (Malin 2020). In some cases, the creators themselves are responsible for changes to content on streaming platforms. Speaking in an interview, the Duffer Brothers, creators of Netflix's *Stranger Things,* impishly confess to "George Lucas-ing the situation" with regards to modifications to visual effects shots in earlier episodes, a reference to the Star Wars creator's penchant for tinkering with his films years and even decades after they were first released. Without going into more specifics about exactly what was changed, the Duffers add, "You do have the physical copies though, the Blu-rays and stuff—you'd have to compare" (Aurthur 2022). While changes to streaming content may well go unnoticed by viewers, the apparent mutability of streaming content is antithetical to the archivist's mission of preserving the cultural record, particularly for films or series that are not released on DVD or Blu-ray, where the only version of record is what is currently on the streaming platform. The ease and speed with which content can disappear in the digital space is concerning for preservation; a *New York Times* article noted that "An episode—or potentially an entire series—can be ordered down, fully removed from streaming and access-restricted in less than 24 hours" (Malin 2020).

Locally managed collections afford libraries a greater degree of preservation control, as they entail a digital source file that is stored and managed through the library's own network. However, this is an area fraught with potential challenges as well. Perpetual access licenses (PAL) for locally managed streaming files typically specify the "life of file" in the license terms. The licenses may also explicitly forbid duplicating files or conversion to other file formats. This introduces the potential for obsolescence, as the file formats used today may not be supported by the video-hosting platforms of tomorrow. Ironically, this mirrors the problems encountered with physical video formats such as VHS. A proactive workaround can be to establish the library's right to convert locally managed files to other formats in the license agreement, although not all content providers may be amenable to such terms.

The rise of streaming can be viewed along a continuum of video format changes, as Lamphere notes, from the earliest films shot on highly flammable nitrate stock to more stable formats, leading to VHS and the home video boom, and then onto the recent transition to digital formats, starting with DVDs and eventually leading to the current epoch of streaming (Lamphere 2020). Every evolutionary stage of video formats has led to a loss of content. The Library of Congress estimates that 70 percent of American silent-era features are completely lost (Pierce 2013). Due to the nature of nitrate film stock, entire studio vaults were lost in fiery conflagrations. In less dramatic fashion, studio vaults from the VHS era may remain in analog formats, left undigitized if the effort to do so is not financially justifiable. It is the nature of video content that titles get left behind with each format iteration, a difficult reality to accept from a preservation standpoint. Regarding this state of affairs, Lamphere goes on to lament "As a librarian with an archivist's heart, it is frustrating to think about the media we have lost and makes me fearful about what we will lose in the future as digital technology continues to develop" (Lamphere 2020, 31).

While the preservation challenges of streaming video as a format are vast and well-known, proposed solutions are in short supply. Since streaming collections are predominantly licensed through short-term access models, the immediate user need is the primary driver for libraries, leaving preservation as an afterthought, if it is thought of at all. Libraries that are engaging in the more permanent building of streaming video collections, often through locally managed holdings, are usually doing so at an individual institutional level and not in coordination with other libraries. Community approaches that pool the collaborative efforts of libraries and publishers, such as the Portico digital repository, have not developed with regard to preserving streaming video. While Portico has to date preserved over 300,000 video files, its primary focus is e-journals and e-books (Portico n.d.). Among libraries, the Academic Library Video Trust (https://www.videotrust.org/about-alvt) has developed as a member-based organization for digitizing video from at-risk formats such as VHS. Member libraries can access streaming versions of content they already hold physically from a shared repository. Another community-based project, the Internet Archive, holds over 10 million videos, with many being archival news footage and short clips similar to YouTube. Like YouTube, it too is subject to copyright uncertainty, as users may upload content that they do not own rights to. A stable preservation model for born-digital video content licensed by libraries has yet to fully emerge.

There has also been little indication that the movement toward making more e-journal and e-book content open access (OA), will carry over to streaming video. Another burgeoning area of openness for libraries is the use of open educational resources (OER) for instructors to freely use and adopt, with the goal of relieving students of the financial burden of purchasing expensive course content such as textbooks. The OER Commons, a repository that makes OER discoverable, actually contains a fair amount of streaming video, mainly in the form of recorded lectures (https://oercommons.org/). With that said, there is no movement toward "flipping" the most widely used streaming video content to OA, as some major publishers have begun to do with e-book and e-journal content. As a newer format, and one that has lucrative potential for distributors, pivots toward OA seem unlikely at this time, particularly since the current access-centered model allows the reselling of the same content year over year, within the walled confines of a proprietary platform. This model, increasingly utilized in e-book publishing, undermines the core of what is considered "ownership" of content, so much so that it has been described as "anti-ownership" (Lamdan et al. 2023). There are strong parallels between the distribution of streaming video and the existential questions casting doubt on the possibility of preservation. After all, how can something be preserved if it is never truly owned to begin with? Taken altogether, the prospects for the preservation of streaming video are less than assured.

Conclusion

It is a final irony that while streaming video continues to grow as a medium, preservation still frequently requires libraries to maintain video content in physical formats. This becomes increasingly difficult as more video content is born digital and acquisition models move away from ownership to access-driven leased models. Opportunities for unique collection building will exist for libraries that can license video content perpetually to be hosted on locally managed platforms, similar to how institutional repositories and online archives are used to showcase digital special collections. Primarily though, libraries will function in the role of access provider, responding to user needs for common titles that are licensed on a short-term basis. While the work of streaming video collection management will continue for libraries, streaming video collections themselves will be highly uniform and largely ephemeral.

References

Andrews, Travis M. September 7, 2022. "Streaming TV is having an Existential Crisis, and Viewers Can Tell." *The Washington Post*. https://www.washingtonpost.com/arts-entertainment/2022/09/07/streaming-tv-changes-crisis/.

Aurthur, Kate. June 24, 2022. "'Stranger Things' Creators Discuss 'George Lucasing' Will's Birthday, Updating Season 4 after it Premiered, and their Not-So-Secret Spinoff Show." *Variety (online)*. https://variety.com/2022/tv/news/stranger-things-season-4-duffer-brothers-spinoff-1235301947/.

Bailey, Lila, Kyle K. Courtney, David Hansen, Mary Minow, Jason Shultz, and Michelle Wu. 2018. "Position Statement on Controlled Digital Lending." Controlled Digital Lending by Libraries. https://controlleddigitallending.org/statement/.

Berman, Judy. February 9, 2023. "The Streaming Wars have Entered their Chaos Era." *Time (online)*. https://time.com/6253697/streaming-wars-disney-chaos-era/.

Colorado Alliance of Research Libraries. n.d. "Member Libraries." Accessed September 25, 2023. https://coalliance.org/member-libraries.

Cooper, Danielle M., Dylan Ruediger, and Makala Skinner. 2022. "Streaming Media Licensing and Purchasing Practices at Academic Libraries: Survey Results." *Ithaka S+R*. https://doi.org/10.18665/sr.316793.

Cooper, Danielle and Roger C. Schonfeld. June 15, 2021. "A Media Empire for the (Public) Library? OverDrive Acquires Kanopy." *The Scholarly Kitchen*. https://scholarlykitchen.sspnet.org/2021/06/15/overdrive-kanopy/.

DiVittorio, Katy, Philip Gaddis, Sommer Browning, Molly Rainard, and Charissa Brammer. 2020. "SILLVR: Streaming Interlibrary Loan Video Resources." *Collaborative*

Librarianship 11(4): 295–312. https://digitalcommons.du.edu/collaborativelibrarianship/vol11/iss4/8/.

DiVittorio, Katy, Philip Gaddis, Rose Nelson, Neil Gilstrap, and Dana Sinclair. 2022. "SILLVR: Streaming Interlibrary Loan Video Resources *and* Accessibility in Multimedia Content: Where Did We Come From, What are We, Where are We Going?" *The Serials Librarian* 82(1–4): 35–9. https://doi.org/10.1080/0361526X.2022.2040234.

Dotson, Daniel S. 2022. "Streaming Videos: Subject Area Usage Trends." *Collection Management* 48(2): 97–113. https://doi.org/10.1080/01462679.2022.2081831.

Foley, Marianne. 2018. "Streaming Video PDA: Brace Yourself, Usage is Coming." *Proceedings of the Charleston Library Conference*. http://dx.doi.org/https://doi.org/10.5703/1288284317017.

Hansen, David R. and Kyle K. Courtney. 2018. "A White Paper on Controlled Digital Lending of Library Books." Controlled Digital Lending by Libraries. https://controlleddigitallending.org/whitepaper/.

Hayes, Dade and Dawn Chmielewski. 2022. *Binge Times: Inside Hollywood's Furious Billion-Dollar Battle to Take Down Netflix.* New York: William Morrow.

Iglesias, Violaine. October 27, 2020. "Video is here. Time to Embrace it." *The Scholarly Kitchen*. https://scholarlykitchen.sspnet.org/2020/10/27/guest-post-video-is-here-time-to-embrace-it/.

Kafka, Peter and Rani Molla. January 5, 2023. "The Streaming Boom is Over." *Vox.com*. https://www.vox.com/recode/2023/1/5/23539590/streaming-losses-netflix-hbo-peter-kafka-media-column.

"Kanopy Launches New and More Affordable Acquisition Models." October 20, 2020. *Video Librarian*. https://videolibrarian.com/articles/news/kanopy-launches-new-and-more-affordable-acquisition-models/.

Kenworthy, Alexander D. 2020. "The Streaming Wars: The Future of Entertainment." *Claremont McKenna College Senior Theses.* https://scholarship.claremont.edu/cmc_theses/2344/.

Koblin, John and Brooks Barnes. May 1, 2023. "Hollywood Writers Go on Strike, Halting Production." *The New York Times.* https://www.nytimes.com/2023/05/01/business/media/hollywood-writers-strike.html.

Lamdan, Sarah, Jason M. Schultz, Michael Weinberg, and Claire Woodcock. 2023. "The Anti-Ownership Ebook Economy: How Publishers and Platforms have Reshaped the Way We Read in the Digital Age." Engelberg Center on Innovation Law & Policy: NYU School of Law. https://www.nyuengelberg.org/outputs/the-anti-ownership-ebook-economy/.

Lamphere, Carly. 2020. "Streaming Media: An Access and Preservation Game Changer." *Online Searcher* 44 (3): 31–4.

Lear, Christian. 2022. "Controlled Digital Lending of Video Resources: Ensuring the Provision of Streaming Access to Video for Pedagogical Purposes in Academic Libraries." *Journal of Copyright in Education and Librarianship* 5 (1): 1–19. https://doi.org/10.17161/jcel.v5i1.14807.

"List of Missing Movies." n.d. *Missingmovies.org.* Accessed September 25, 2023. https://missingmovies.org/list-of-missing-movies/.

Malin, Sean. August 17, 2020. "What Actually Happens When a TV Episode Gets Pulled?" *The New York Times*. https://www.nytimes.com/2020/08/17/arts/television/pulled-episodes-blackface.html.

Mason, Katherine F. 2022. "Managing Streaming Video License Renewals." *Serials Review* 48 (3–4): 206–7. https://doi.org/10.1080/00987913.2022.2097849.

"The 'Missing Movies' Manifesto." n.d. *Missingmovies.org.* Accessed September 25, 2023. https://missingmovies.org/the-missing-movies-manifesto/.

Nathanson, Michael. September 14, 2022. "How the Streaming Wars Will Alter the Media Landscape." *Yale Insights.* https://insights.som.yale.edu/insights/how-the-streaming-wars-will-alter-the-media-landscape.

Pallotta, Frank. August 11, 2022. "The Streaming Wars are Over." *CNN Business*. https://www.cnn.com/2022/08/11/media/streaming-disney-netflix/index.html.

Pierce, David. September 2013. "The Survival of American Silent Feature Films: 1912–1929." Washington, DC: Council on Library and Information Resources and The Library of Congress. https://www.clir.org/wp-content/uploads/sites/6/pub158.pdf.

Portico. n.d. "Facts and Figures." Accessed September 25, 2023. https://www.portico.org/coverage/facts-and-figures/.

Rodgers, Wendy. 2018. "Buy, Borrow, or Steal? Film Access for Film Studies Students." *College & Research Libraries* 79 (4): 568–91. https://crl.acrl.org/index.php/crl/article/view/16730.

Rottenberg, Josh. June 13, 2023. "Streaming's Most Wanted: Why it's So Hard to Find these Iconic Film and TV Titles." *Los Angeles Times.* https://www.latimes.com/entertainment-arts/story/2023-06-13/moonlighting-dawn-of-the-dead-not-streaming-explained.

Serrano, Amauri and Michael Fernandez. 2023. "Impacts of COVID-19: Toward a Streaming-Preferred Video Collection Policy." *Collection Management* 48 (1): 5–21. https://doi.org/10.1080/01462679.2022.2124392.

Statista. July 31, 2023. "Forecast of Video-on-Demand Users by Segment in the United States from 2019 to 2027 (in million)." Chart. Accessed September 26, 2023. https://www.statista.com/forecasts/459423/video-on-demand-users-in-the-united-states-forecast.

Tanasse, Gisèle. 2021. "Implementing and Managing Streaming Media Services in Academic Libraries." ACRL/Choice paper. https://www.choice360.org/wp-content/uploads/2021/10/Choice_Implementing_and_Managing_Streaming_Media_Services_in_Academic_Libraries_lo_res_100721.pdf.

Urban, Sandra Gall. 2022. "Evaluating Kanopy Access Models in Academic Libraries: Balancing Demand and Budget Constraints." *Collection Management* 47 (1): 37–48. https://doi.org/10.1080/01462679.2021.1940410.

Appendix A

Collection Development Statement: Video Yale University Library (Public)

Overview

Yale University Library's video collection development strategy is motivated by a desire to meet the teaching, learning, and research needs of all Yale students and faculty. The extensive collection of physical media and a growing number of streaming videos expands and strengthens the library's robust general collections while also deepening the collections in key research areas. The goal is to provide the Yale community access to a large, diverse, and long-lasting media collection.

Yale University Library's video collection includes short and feature-length fiction features and shorts, documentaries, animated films, theatrical adaptations, television programs, experimental films, and video art available streaming, and on DVD, Blu-ray, VHS videotapes, and LaserDiscs (the latter two formats are no longer collected). The Yale Film Archive (https://web.library.yale.edu/film) is the primary video research service point and provides access to video playback equipment and houses new video acquisitions as well as the largest circulating video collection outside of the Library Shelving Facility (LSF). Documentaries and subject- and genre-specific content are also found in the Haas Family Arts Library and the Gilmore Music Library. In addition to physical media, the library also provides access to an ever-growing number of streaming videos that are locally managed or hosted on vendor platforms.

History

Prior to the Film Study Center, now known as the Yale Film Archive, joining the library in 2017, the library's collection consisted primarily of documentaries, a select number of fiction films, and films to support the arts and performing arts. The collection was meant to complement the Film Study Center's holdings of short and feature-length fiction films, documentaries, television series, and theatrical productions on

DVD and Blu-ray. After the two physical collections merged in 2017, the Film Study Center's holdings were cataloged and added to the library's discovery systems, video selection acquisition responsibilities were redefined, and the library revised its overall video collecting strategy.

Beginning in the early 2000s, subject librarians began to acquire vendor-hosted subject- and genre-specific streaming video and mixed media collections and continue to do so today. In recent years, the library has explored various streaming video purchasing models including the perpetual purchase of collections and individual film titles, patron-driven acquisitions (PDA), evidence-based acquisitions (EBA), and subscriptions to multi-subject and multi-genre streaming collections.

Collection Highlights

- Japanese films; German films; British films; films by African American filmmakers and/or on African American subjects; and films by LGBTQ filmmakers and/or on LGBTQ subjects
- Premier collections on DVD and Blu-ray such as the Criterion Collection
- British Film Institute's decennial "Greatest 250 films of all time" polls
- 3-D Blu-rays for screening at the Yale Film Archive
- Streaming documentaries from academic distributors

Expanding Areas of Focus

- World cinema is underrepresented in Library video collections from Latin America, Spain, Portugal, Eastern Europe, the Middle East, South and Southeast Asia, Africa, and East Asia
- US Diversity, equity, and inclusion (DEI) focus—African American Studies, Asian American Studies, Native American Studies, Latinx Studies, Women and Gender Studies in all film categories
- Streaming video in all areas and categories of film—subscriptions and perpetual purchases

Academic Departments and Programs Supported

Video collection supports departments and programs throughout the University.

Selection Guidelines and Policies

Selection Responsibilities

Selection responsibilities for video are distributed across libraries and are based on format, subject area, and curricular and research needs.

The Yale Film Archive has the primary responsibility for the selection and acquisition of film. YFA collects

- motion pictures and TV series to support known campus research and teaching interests
- works with unique places in film history
- works from major filmmakers and filmmakers from underrepresented areas
- significant works as determined by major film critics' reviews, annual awards from major film critics' societies and associations, annual awards (including nominees), from major organizations and film festivals, British Film Institute (Sight and Sound) polls
- works from eras, movements, genres, or national cinemas underrepresented in the library collections
- feature films and shorts from North America and Western Europe (primary responsibility)

General Selection Responsibilities

- Course Reserves
  - Yale Film Archive—all academic departments except for courses taught in the Medical, Divinity, and Law schools
  - Medical, Divinity, and Law libraries support courses in respective schools
- Fiction Films and Shorts
  - Area Studies librarians—foreign language films from Spain, Portugal, Latin America, the Caribbean, East, Central, South and Southeast Asia, Eastern Europe and Russia, and the Near East
  - Yale Film Archive in accordance with department selection areas
  - Yale Film Archive in collaboration with the appropriate subject librarian—requests for films from international distributors, expensive films, or individual requests for multiple films
  - Other librarians—selectively and by request
- Documentaries
  - Subject librarians in accordance with subject-specific collecting criteria (primary responsibility)
  - Yale Film Archive in accordance with selection areas
- Theatrical Productions
  - Music and performing arts librarians in accordance with subject-specific collection criteria (include link to statements)
- Multidisciplinary Streaming Video Collections
  - Managed by the library's Collection Development department in collaboration with the Yale Film Archive (e.g., Kanopy)
- Profession-specific Video
  - subject librarians in accordance with subject-specific collecting criteria

Formats Collected

- DVD
- Blu-ray
- Streaming
- Databases
  - Collections—perpetual access
  - Collections—limited license period (e.g., Criterion Collection on Kanopy)
  - Individual videos—perpetual access
  - Individual videos—limited license period (only for curricular purposes or specific research requests when a physical format is not available or appropriate)

Streaming preferred

- Course reserves
- Replacement of heavily used videos owned in physical format, when the video is available for perpetual access on a vendor platform with no hosting fee
- Films from academic distributors available for perpetual purchase for which the physical and streaming versions have comparable pricing

Languages Collected

The library acquires videos in English or foreign languages with subtitles. Foreign language productions without English subtitles are acquired on a selective basis, usually in response to a curricular request from faculty.

Other Selection Considerations

Duplication

In general, the library does not duplicate streaming and physical media purchases, nor acquire more than two copies of the same DVD or Blu-ray.

Criteria to purchase multiple copies of the same DVD or Blu-ray

- The film is used for course reserve and no streaming version is available
- The film is regularly used for curricular and research needs and no streaming version is available
- Significant collections (e.g., Criterion Collection)

Criteria to purchase a DVD/Blu-ray copy for a film that has been licensed streaming for perpetual use*

- DVD/Blu-ray includes extras (e.g., behind-the-scenes documentaries, interviews, scene analyses, booklets, etc.)
- DVD/Blu-ray has various versions of the film
- DVD/Blu-ray includes various audio tracks, including choices of languages, and commentaries by directors, writers, cinematographers, actors, etc.
- DVD/Blu-ray includes subtitles in various languages or the option not to view subtitles
- DVD/Blu-ray has better picture quality, especially when shown on a large screen
- DVD/Blu-ray is necessary for close analysis, especially shot-by-shot
- DVD/Blu-ray is utilized for clip capture when not available on the streaming provider platform

*DVDs/Blu-rays are selectively acquired for films that are not available streaming for perpetual purchase.

Streaming Availability: Consumer-Facing Platforms

If the streaming version of a film is only available on a consumer-facing platform, the library will purchase the DVD when available. If a DVD is not available, the user must pay for personal access via a subscription or video rental.

Streaming Selection Considerations

Purchase models and licensing selection preferences

- Purchase or licensing must allow for educational, instructional/teaching, or classroom use
- Perpetual streaming rights are acquired when available at a reasonable price
- Availability of digital files for archival purposes for perpetually acquired video including vendor-hosted content
- Subscriptions to streaming video collections are preferred
- Individual videos with term licenses are licensed only for curricular or research needs
- Videos sourced directly from copyright holders are licensed selectively and primarily for curricular purposes

Accessibility and vendor platform preferences

- Closed-captioning or caption files
- Video transcripts—searchable, printable, or interactive
- Player keyboard controls
- Audio description
- Range of resolutions, applicable to varying broadband availability

- Reliable player, compatible with most browsers, and allows for embedding/linking into websites/course management systems

Access criteria

- No restrictions on remote access
- Unlimited simultaneous users
- Minimal or no maintenance or access fees for purchased content hosted on vendor platforms
- Distributors and platforms that support user privacy

Gifts

In line with the Library Gift Policy, offers of rare materials and audio and visual material (including phonograph records, CDs, VHS tapes, and DVDs) are generally *not* considered to be rare or unique, and therefore rarely accepted.

Public Performance Rights

Public Performance Rights (PPR) are considered during the selection process and are acquired as part of the purchase of an item. Library catalog records for physical formats note the availability of PPR.

PPR are not required for face-to-face teaching, individual and home viewing, or viewing in small groups consisting of members of the Yale University Community.

PPR are required for all screenings of copyrighted media to audiences outside of the regular curriculum. The library does not acquire PPR for campus events.

Exclusions

- VHS tapes, Beta tapes, LaserDiscs, or other legacy video formats
- Language instruction videos, unless included in multidisciplinary streaming databases or collections
- PPR for campus events

Streaming Video Selection Guidelines Yale University Library (Internal)

I. Purpose

The streaming video market is continuously changing, as are user needs and requests. Streaming video presents unique problems within an educational setting such as management of user expectations, sourcing content, licensing, and access.

These guidelines are intended for subject or collection management librarians who would like to acquire or license streaming video. They provide general criteria and procedures for the selection, acquisition, and access to streaming video content. However, because of the uniqueness of streaming video requests, these guidelines are by no means comprehensive. Unless otherwise noted in these guidelines, please contact your departmental collection manager with questions.

Table A.1 Individual Videos

Option	Preference	Example
Perpetual streaming rights or digital site license	High	Purchase of individual film files directly from a content provider with perpetual streaming rights. Hosting fees may apply for films purchased in perpetuity and hosted on a vendor's platform.
DVD + (perpetual) streaming rights	High	Women Make Movies documentary (DVD + streaming rights for the life of file or just streaming w/o DVD).
Limited streaming license	Medium (only for course-related and research purposes)	Individual video licensed for a limited period and hosted on a platform with institutional access (e.g., Kanopy or Swank).
Sourced content directly from the rights holder	Low (i.e., faculty request and no other options available)	Find, request, and purchase film files directly from the filmmaker or distributor.
DVD only + direct users to stream independently	Low (streaming requested but no institutional streaming option available)	DVD purchase and recommend individual paid access via a consumer-facing platform. Librarians may direct users to JustWatch (https://www.justwatch.com/us) to determine which services provide the needed content.

Table A.2 Collections and Packages

Option	Preference	Example
One-time acquisition of streaming video collection	High	Docuseek2—one-time video collection purchase with the option to host on vendor platform for a small fee or locally hosted.
Subscribed video package/collection	Medium	Films on Demand—subscription to an entire collection.
Patron-Driven Acquisitions or Evidence-Based Acquisitions	Low	Most subscription video collections offer some form of PDA or EBA. Kanopy offers PDA for the entire collection or supplier collections within the larger collection. The library prefers not to use PDA and EBA models.

II. Acquisition and Licensing Options

Other Considerations

- PPR are preferable and should be considered during selection. The acquisition of PPR is determined by the librarian considering cost and expected use.

III. Patron Requests

- Purchase/license all faculty-requested videos and anything used for course-related and research purposes.
- Course Reserves: ask instructors to submit course reserve requests via Canvas. Yale Film Archive fills streaming video course reserves.
- Research Requests: source video using selection guidelines. If no institutional purchase option is available, librarians may refer users to a commercial platform or provide links to freely available content with the disclaimer that we cannot guarantee the legitimacy or longevity of free online content.
- Personal Use Requests (i.e., not for research): librarians may refer students/staff to a commercial platform or share links to free content highlighting that we are only licensing individual videos for course-related and research purposes. Librarians should inform students/staff that we cannot guarantee legitimacy or longevity of free online content.

IV. Cost

- General: The librarian determines the reasonable prices and how to cover the costs.

- Multidisciplinary Subscription Streaming Video Collection: Librarians should consult their departmental collection manager for possible cost-shares or other modes of funding.

V. Platform Features and Picture Quality

Library prefers:

- Provision of machine-readable cataloging (MARC) records (for video collections)
- Provision of usage statistics in compliance with the current COUNTER Code of Practice
- Accessibility features
  - closed captions
  - video transcripts—searchable, printable, or interactive (dialog text displayed next to the video that allows the user to click on a word and go straight to that spot in the video)
  - player keyboard controls—play, pause, etc., with keyboard and not mouse
  - audio description—an extra audio soundtrack that describes what is happening
- Range of resolutions, applicable to varying broadband availability
- Reliable player, compatible with most browsers, and allows for embedding/linking into websites/course management systems

VI. Language

- The library primarily acquires video in English or a foreign language with subtitles. Foreign language productions without English subtitles will be acquired on a selective basis, usually in response to a specific curricular request from faculty.
  - Selectors working with non-English resources will assist in vendor communication as needed.

VII. Copyright and Licensing

- Streaming video purchase or licensing must allow for educational, instructional/teaching, or classroom use.
- PPR are needed for any event that is open to the general public. Unless PPR are expressly stated in the license, the library must secure permission from the rights holder.
- Some academic streaming video collections offer public performance rights (e.g., Docuseek2, Digitalia Film Library). Other streaming video providers allow group viewings by Yale-Authorized Users in and out of the classroom (e.g., Kanopy). Other providers only allow teaching and research-related use (e.g., Swank).

- Consistent with the University's commitment to complying with all applicable intellectual property laws and its expectation that all university community members respect the copyright rights of others, the Library may not digitize copyrighted full film/video content without express written permission from the rights holder.
  - Faculty may work with the Poorvu Center for Teaching and Learning to excerpt short film clips in accordance with TEACH Act exemption guidelines.
- All copyright and licensing questions should be directed to the Library's License Review Steering Committee.

Library prefers:

- a perpetual digital site license (DSL)
- established vendor with which the library has a current license on file
- video hosted on the vendor platform

VIII. Access and Availability

Library prefers:

- perpetual access to purchased content
- digital files to purchased content even if hosted on a vendor platform
- close captioning or caption files whenever possible

For vendor-hosted content, the library prefers:

- minimal or no maintenance and/or access fees for purchased content hosted on a vendor platform
- stable links
- no advertisements on the interface or embedded in the content
- ability to add Yale University Library branding
- content can be streamed via EZproxy for remote access to Yale Authorized Users
- no restrictions on remote access (e.g., students may only watch films while in the classroom and not from residence halls)
- distributors and platforms that support the privacy of our users (e.g., do not collect personally identifiable user data)
- Librarians are expected to provide any vendor-supplied information when placing an order
- Vendor provides access to media file (non-institutional platform)
  - Type of authentication (e.g., IP address, single sign-on, username/password)
  - Is remote access permissible and if so, how (e.g., VPN, EZproxy)?
  - Number of simultaneous users
- Vendor provides file that will be hosted locally
  - Preferred format: MP4 files with H.264 codec

- Preferred file size: 3 GB or less
- Preferred format for caption file: SRT files
- Preferred file delivery method: download link via DropBox or similar file hosting site

IX. Access and Discovery

- The discovery of streaming video at YUL is dependent on current internal cataloging practices.
- As a result, librarians are expected to
  - appropriately search for current access to streaming video content at YUL before placing orders.
  - choose the appropriate acquisition model and licensing option for the intended use (e.g., discoverability in the library catalog is not as important if a video is needed for course reserves).
  - Ask vendors if they supply MARC records. If no MARC records are provided and discoverability is necessary, Yale staff will have to perform original cataloging for search/discovery purposes.

Streaming video access and discovery at YUL

- Streaming video collections (subscribed and purchased) and individually purchased streaming videos (with perpetual access) will be cataloged for inclusion in the library catalog.
  - The ability to include individual streaming video titles in the catalog may be limited by inconsistent or low-quality vendor-provided metadata.
- Streaming video titles licensed individually for limited terms will not be cataloged.
- Streaming video collections will be activated at the Database level in the library's 360 Knowledgebase.
- Streaming video collections may be added to the Databases A–Z list at the request of the librarian.
- Some of streaming video providers add new content to their collections; if a desired title is not found in the catalog, consult the streaming video guide at https://guides.library.yale.edu/streamingvideo for the most current list of collections.

Appendix B

Streaming Video Model License

Note: This template was drafted by Sandra Enimil, Amauri Serrano, and Michael Fernandez.

THIS STREAMING VIDEO/DIGITAL MATERIALS LICENSE AGREEMENT is made effective as of **[date]** (the "Effective Date") between **[Name of Publisher], [Address of Publisher, City of Publisher, State of Publisher, Postal Code of Publisher]** (heretofore referred to as "Licensor") and **[Name of Library], [Address of Library, City of Library, State of Library, Postal Code of Publisher]** (heretofore referred to as "Licensee").

The materials that are the subject of this Agreement shall consist of certain electronic files or information published or otherwise made available to Licensee by or through Licensor, specifically identified as follows (hereinafter referred to as the "Licensed Materials"): **[title and description of materials].**

[Use one of the following clauses depending on whether a short-term license or perpetual/life of file license is being negotiated.]

Perpetual License Term. The term of this Agreement shall be for the life of the file or duration of copyright term, whichever is longer.

OR

Limited License Term. The term of this Agreement shall begin on **[start date]** and terminate on **[end date].** This Agreement may be renewed or extended upon the mutual agreement of the parties.

Fee. Licensee shall pay Licensor **[license fee].**

Digital File(s).

[Use one of the following clauses depending on whether the file delivery method is a vendor-provided file or if the library is digitizing from a pre-owned source, e.g., a DVD held in the collection]

Licensor shall provide a **[file format]** delivered via **[delivery method].**

OR

Licensee is responsible for digitizing the Licensed Materials.

Licensee shall have the right to transcode the content into the streaming format of the Licensee's choice and to load the digital files onto the Licensee's digital network for streaming to **[Name of Library]** Authorized Users. Licensed Materials may be duplicated for the purpose of a backup copy, and Licensee is authorized to convert Licensed Materials using any medium, format, or technology now in existence or hereinafter devised, including the internet, for the uses permitted herein.

"Authorized Users" are full- and part-time students and employees and contractors (including faculty, staff, and affiliated researchers and users) and other valid ID holders accepted by **[Name of Library]**.

Permitted Access and Use.

- Licensor hereby grants to Licensee the right to make the Licensed Materials available to Authorized Users for educational, scholarly, and research purposes, and any other non-commercial use in support of Licensee's mission, as well as hold non-theatrical screenings where admission is not charged.
- Licensed Material(s) are for synchronous (live) or asynchronous streaming (multiple users, video on demand format) only and content therein may not be copied to any other format or used for any purpose not specified in this Agreement. Licensee is permitted to host and stream the Licensed Materials from a secure server with password-protected access for Authorized Users.

Prohibited Uses. Licensee will not mount or distribute any element of the Licensed Materials on any electronic network accessible to parties who are not Authorized Users, including without limitation the internet and the World Wide Web; use all or any part of the Licensed Materials for commercial use; or copy, transmit, modify, distribute, sell, or create derivative works of the Licensed Materials; in all of the foregoing cases, except as permitted by United States copyright law, or as otherwise permitted herein.

Accessibility. Licensor shall deliver transcripts and caption files when available. Licensee is authorized to add pop-on or closed-captioning to the digital file(s). Nothing in this Agreement shall limit the Licensee or any end user from making lawful, non-infringing uses to facilitate access to the Licensed Materials by users who have disabilities as defined under the Americans with Disabilities Act, as amended. For the avoidance of doubt, the Licensor authorizes such uses.

Representations and Warranties. Each party represents and warrants that it has full power and authority to enter into and perform its obligations under this Agreement, and that the person who has executed this Agreement on behalf of such party has the authority to bind such party.

Relationship of the parties. Nothing contained in this Agreement will be deemed to create an agency, joint venture, partnership, or other similar relationship between the parties. Neither party has the authority to bind the other party in any way, nor will either party represent that it has such authority.

Miscellaneous. This Agreement constitutes the entire agreement of the parties and supersedes any prior written or oral agreement or understandings between them with regard to the Licensed Materials, including without limitation (if applicable), Licensor's online/website Terms and Conditions and/or any shrink wrap contracts packaged with physical copies of the Licensed Materials. This Agreement cannot be changed or terminated orally but only in writing signed by both parties. Neither party shall assign this Agreement and any purported assignment to the contrary shall be null and void. This Agreement shall be governed by the laws of **[preferred judicial venue of your library]**, without reference to its choice of law principles. All disputes hereunder shall be resolved in the applicable state or federal courts of **[preferred judicial venue of your library]**.

IN WITNESS WHEREOF, the parties have executed this Agreement by their respective, duly authorized representatives as of the date first above written.

LICENSOR:
BY: ______________________________________ DATE: _______________
Signature of Authorized Signatory of Licensor
Print Name:
Title:
Address:
Telephone No.:
E-mail:

LICENSEE:
BY: ______________________________________ DATE: _______________
Signature of Authorized Signatory of Licensee
Print Name:
Title:
Address:
Telephone No.:
E-mail:

Appendix C
Cataloging Template for Streaming Video

Note: This template was drafted by Rowena Griem, Cataloging Librarian at Yale University Library. It is partially adapted from OLAC's *Best Practices for Cataloging Streaming Media.*

For additional information, see *Best Practices for Cataloging Streaming Media Using RDA and MARC21* (OLAC).

Fixed Fields		
Leader		Type of Record (Leader/06): g = Projected Medium Bibliographic Level (Leader/07): • m = Monograph/item (for single titles/MPMs) • i = integrating resources (for ongoing collections)
006	Form of Material (006/00): m = Computer File	Form of Item (006/06): o = Online resource Type of File (006/09): c = Representational
007 #1 (One 007 field for each category of material)	Category of Material (007/00): v = Video Recording	Specific Material Designation (007/01): z = Other Color (007/03): • b = Black and white • c = Multicolored • m = Mixed • u = Unknown Format (007/04): z = Other Sound on Medium (007/05): a = Sound on medium (unless the film is silent) Medium for Sound (007/06): z = Other Dimensions (007/07): u = Unknown Playback Channels (007/08): u = Unknown

(Continued)

Fixed Fields		
007 #2	Category of Material (007/00): c = Computer File	Specific Material Designation (007/01): r = Remote Color (007/03): • b = Black and white • c = Multicolored • m = Mixed • u = Unknown Dimension (007/04): n = Not applicable Sound on medium (007/05): a = Sound on medium (unless the film is silent)
008		Form of Item (008/29): o = Online

Variable Fields

Field	Field Text	Notes
041 Language Code	041 0/ ǂa [Language code(s)] ǂj [Language code of subtitles]	*First indicator = 1 for resources with subtitles in translation*
043 Geographic Area Code	043 // ǂa [Geographic code(s)]	
046 Special Coded Dates	046 // ǂk [Beginning or single date created] ǂl [Ending date]	
130 Main Entry—Uniform Entry	130 x/ ǂa [Uniform title]	
245 Title Statement	245 xx ǂa [Title] = ǂb [Parallel title] / ǂc [Statement of responsibility]	*Only include prominent info in statement of responsibility; most should go in 508 field*
246 Varying Form of Title	246 3/ ǂa [Variant title, including Parallel title]	
257 Country of Producing Entity	257 // ǂa [Country of Production Company] ǂ2 naf	

(Continued)

Variable Fields		
Field	**Field Text**	**Notes**
264 Publication Information	264 /1 ǂa [Location of publisher] : ǂb [Publisher], ǂc [Publication date]	*Description of original release, not video recordings*
300 Physical Description	300 // ǂa 1 online resource (video files)	*For collections*
	• 300 // ǂa 1 online resource (1 video file (approximately 115 min.)): ǂb sound, color • 300 // ǂa 1 online resource (2 video files (25 min., 45 sec. each)) : ǂb silent, black and white	*For single titles*
306 Playing Time	306 // ǂa [coded duration in hours, minutes, and seconds]	e.g., 010814 = 1 hr, 8 min., 14 sec.
336 Content Type	336 // ǂa two-dimensional moving image ǂb tdi ǂ2 rdacontent	
337 Media Type	337 // ǂa computer ǂb c ǂ2 rdamedia	
338 Carrier Type	338 // ǂa online resource ǂb cr ǂ2 rdacarrier	
380 Form of Work	380 // ǂa [Predominant genre] ǂ2 [Source of term, e.g., ‡2 lcgft]	e.g., Documentary films, Feature films, Filmed performances, Short films, Television programs
500 General Note	500 // ǂa Originally produced as a motion picture in [year of original release]	
506 Restrictions on Access Note	506 // ǂa Access restricted by licensing agreement	

(Continued)

Variable Fields		
Field	**Field Text**	**Notes**
508 Creation/Production Credits	508 // ǂa [Composer; Director; Editor; Producer; Screenwriter]	*Separate components with [space]; [space]*
511 Participant/Performer Note	511 / ǂa [Actor; Conductor; Dancer; Narrator; Performer]	*First indicator "1" for cast, "0" for other*
518 Date/Time/Place of Event	518 // ǂa [For filmed performance or event]	
520 Summary, etc.	520 // ǂa [Summary]	
540 Terms Governing Use and Reproduction Note	540 // ǂa Public performance rights included ǂ5 CtY 540 // ǂa Public performance rights included ǂg [expiration date of use right, YYYY-MM-DD] ǂ5 CtY	
546 Language Note	546 // ǂa [Language Note, e.g., in French with English subtitles; In English with optional English captions]	
588 Source of Description Note	588 // ǂa Description based on [physical format] version record	*If record for e-version is based on record for physical format*
	588 // ǂa Description based on online resource; title from title screen ([Provider], viewed [Date])	*If born digital*
6xx Subject Access Entries	6xx x0 ǂa [Subject heading]	*Include* ǂv Drama *form subdivision for dramatizations*

(Continued)

Variable Fields		
Field	**Field Text**	**Notes**
655 Genre/Form	655 /7 ‡a [Genre/form heading(s)]. ‡2 lcgft	e.g., Documentary films, Documentary-style films, Experimental films, Fiction films, Film adaptations, Filmed performances, Television programs
655 Genre/Form	655 /7 ‡a Films for the hearing impaired. ‡2 lcgft	*Add for all films with captions*
	655 /7 ‡a Films for the hearing impaired. ‡2 lcgft	*Add for all films with captions*
7xx Added Entry—Responsible Entity	700 or 710 x/ ‡a [Responsible Entity], ‡e [Relationship designator]	e.g., composer, editor of moving image work, film director, film producer, on screen participant, performer, production company
776 Additional Physical Form Entry	776 08 ‡i [Physical format] version ‡a [Title] ‡z [ISBN] ‡w [OCLC number]	*Include if record for e-version is based on record for a physical format*
856 Electronic Locations/Access	856 40 ‡y Streaming video ‡u [URL]	

Appendix D

Educational Streaming Video Providers

Large Streaming Providers

Large streaming vendors act as aggregators of video content from various distributors and producers. They offer a variety of licensing options and host content on their platforms.

Provider: Alexander Street
https://alexanderstreet.com/
Content: feature films and shorts, documentaries, theatrical productions, television series, newsreels
Acquisition Model: subscription with evidence-based acquisitions (EBA; Academic Video Online (AVON)—Build by Choice), term license (1 and 3 years), perpetual access licenses (PAL; single title and collection)
Licensing Options: 1 year, 3 years, perpetual (single title and collection)
Public Performance Rights: Yes, excluding Film Platform and National Theatre titles
Notable Producers: Sony Picture Classics, PBS

Provider: Bloomsbury Video Library
https://www.bloomsburyvideolibrary.com/
Content: theatrical performances, experimental and feature films, documentaries, instructional videos
Licensing Options: subscription (Arts and Humanities Collection or smaller thematic collections), PAL
Public Performance Rights: No

Provider: Criterion Pictures USA
https://www.criterionpicusa.com/
Content: feature films, documentaries
Licensing Options: subscription, term license (1 and 3 years)
Public Performance Rights: No
Notable Producers: 20th Century Fox, Searchlight Pictures

Provider: Digitalia Film Library
https://www.digitaliafilmlibrary.com/
Content: international feature films, documentaries, television series
Licensing Options: subscription(complete library, collections, title-by-title)
Public Performance Rights: Yes

Provider: Docuseek
https://docuseek2.com/
Content: documentaries
Licensing Options: subscription, EBA, DSL for life of file, or provider hosted with platform fee. Flat platform fee for all content purchased
Public Performance Rights: Yes
Notable Producers: Bullfrog, Icarus Films, Film Movement, First Run, Pragda

Provider: Films on Demand (Infobase)
https://infobase.com/products/master-academic-collection/
Content: feature films, documentaries, television series, archival films, newsreels
Licensing Options: subscription (academic package, subject collections, and world cinema collection), term license (1 and 3 years), PAL
Public Performance Rights: Available for some films
Notable Producers: Films for the Humanities, California Newsreel, HBO

Provider: Hoopla
https://www.hoopladigital.com/
Content: feature films, documentaries, television series,
Licensing Options: subscription
Public Performance Rights: No
Notable Producers: A24, Disney Studios, Lionsgate, MGM, Paramount Pictures

Provider: Kanopy
https://lib.kanopy.com/
Content: feature films, documentaries, theatrical productions, television series
Licensing Options: subscription, term license (1 and 3 years), PDA, mediated PDA, Smart PDA (curated PDA), PAL + hosting fee, pay-per-use (PPU)
Public Performance Rights: available for some films (filter available on the platform)
Notable Producers/Suppliers: Criterion Collection, Kino Lorber, DEFA, Samuel Goldwyn

Provider: Swank
https://www.swank.com/
Content: feature films, documentaries, television series

Licensing Options: subscription (Top 1000, Flex, customized collection), DDA, mediated DDA, term license (1 and 3 years)
Public Performance Rights: No
Notable Producers: Columbia Pictures, Universal Studios, Paramount Pictures, Warner Bros.

Independent Distributors

Many independent distributors make their content available on streaming vendor platforms either as term licenses or PALs, but also have titles that must be purchased directly from the distributor and locally hosted.

Distributor: ArtMattan Films
https://artmattanfilms.com/
Content: feature films and documentaries that focus on the experience of people of color all over the world, especially Africa, the Caribbean, North and South America, and Europe
Acquisition Models: DVD, DSL for life of file
Public Performance Rights: Yes
Vendor Platform: Kanopy

Distributor: Bullfrog
https://www.bullfrogfilms.com/
Content: documentaries on environmental and related social justice
Acquisition Models: DVD, DVD + DSL (3 years), DSL for life of file
Public Performance Rights: Yes
Vendor Platform: Docuseek

Distributor: California Newsreel
https://newsreel.org/
Content: documentaries with a focus on social justice especially the African American and African experience
Acquisition Models: DVD, DSL (1 and 3 years)
Public Performance Rights: Yes (only DVD)
Vendor Platform: Alexander Street, Films on Demand, Kanopy

Distributor: Cinema Guild
http://www.cinemaguild.com/
Content: documentaries
Acquisition Models: DVD, DSL for life of file DVD + DSL for life of file
Public Performance Rights: Yes
Vendor Platform: Kanopy

Distributor: Collective Eye
https://www.collectiveeye.org/
Content: documentaries
Acquisition Models: DSL, DVD + DSL for life of file
Public Performance Rights: Yes
Vendor Platform: Alexander Street, Kanopy

Distributor: Documentary Educational Resources (DER)
https://www.der.org/
Content: ethnographic documentaries
Acquisition Models: DVD, DSL for life of file, DVD + DSL for life of file
Public Performance Rights: Available by request
Vendor Platform: Alexander Street, Kanopy

Distributor: Film Movement
https://www.filmmovement.com/
Content: feature films and documentaries
Acquisition Models: DVD, DVD + DSL for life of file
Public Performance Rights: Yes
Vendor Platform: Alexander Street, Films on Demand

Distributor: First Run Features
https://www.firstrunfeatures.com/
Content: documentaries
Acquisition Models: DVD, DSL for life of file; Provider hosted-term license (1 and 3 years) and PAL
Public Performance Rights: Yes
Vendor Platform: Alexander Street, Films on Demand, Docuseek, Kanopy

Distributor: Frameline
https://www.frameline.org/
Content: LGBTQ+ documentaries and animated films
Acquisition Models: DVD, DSL for life of file
Public Performance Rights: Yes
Vendor Platform: Alexander Street, Kanopy

Distributor: Good Docs
https://gooddocs.net/
Content: documentaries
Acquisition Models: DVD, distributor hosted term license (14 days), DSL for life of file, DVD + DSL for life of file
Public Performance Rights: Available
Vendor Platform: Docuseek

Distributor: Grasshopper Film
https://grasshopperfilm.com/
Content: documentaries
Acquisition Models: DVD, DSL for life of file, DVD + DSL for life of file
Public Performance Rights: Yes
Vendor Platform: Alexander Street, Kanopy, ProjectR

Distributor: Icarus Films (Fanlight Collection and dGenerate Films)
https://icarusfilms.com/
Content: documentaries
Acquisition Models: DVD, DSL for life of file
Public Performance Rights: Yes
Vendor Platform: Docuseek

Distributor: KimStim
https://www.kimstim.com/
Content: international feature films and documentaries
Acquisition Models: DVD, DVD + DSL for life of file
Public Performance Rights: Yes
Vendor Platform: Docuseek, Kanopy

Distributor: Kino Lorber
https://kinolorberedu.com/
Content: feature films and documentaries
Acquisition Models: DVD, DVD + PPR, DSL for life of file, DSL for life of file + PPR
Public Performance Rights: Yes
Vendor Platform: Kanopy

Distributor: New Day Films
https://www.newday.com/
Content: documentaries
Acquisition Models: DVD, term license (14 days; 1, 3, and 5 years), DSL for life of file
Public Performance Rights: Yes
Vendor Platform: Kanopy, New Day Digital

Distributor: Passion River Films
https://www.passionriver.com/
Content: feature films and documentaries
Acquisition Models: DVD + PPR, DSL (1 and 3 years), DSL life of file
Public Performance Rights: Available
Vendor Hosting: Alexander Street

Distributor: Pragda
https://pragda.com/
Content: Latinx, Latin American, and Spanish films in all genres
Acquisition Models: subscription (all and individual collections), term licenses (1 and 3 year), DSL for life of file
Public Performance Rights: Yes
Vendor Platform: Docuseek, Pragda STREAM (Pragda STREAM is part of Docuseek, but the distributor has its own homepage)

Distributor: Third World Newsreel
https://www.twn.org/
Content: social change films of all genres
Acquisition Models: DVD, DSL (1, 3 year; life of file)
Public Performance Rights: Available
Vendor Platform: Alexander Street

Distributor: Video Data Bank
https://www.vdb.org/
Description: video and media art
Acquisition Models: DVD/Blu-ray, DSL (1 and 5 year), DSL for life of file, DSL + DVD
Public Performance Rights: Yes
Vendor Platform: Kanopy

Distributor: Video Project
https://www.videoproject.org/
Description: documentaries
Acquisition Models: DVD + DSL for life of file
Public Performance Rights: Yes
Vendor Platform: Alexander Street, Kanopy

Distributor: Women Make Movies
https://www.wmm.com/
Description: documentaries by and about women
Acquisition Models: DVD, DSL for life of file, DVD + DSL for life of file
Public Performance Rights: Yes
Vendor Platform: Docuseek, Kanopy

Discipline or Profession-Specific Providers

This is not a comprehensive list of discipline or profession-specific providers, and it is based on the most commonly used streaming vendors in ITHAKA S+R's 2021 Streaming Media Licensing and Purchasing Practices at Academic Libraries survey.

Provider: Digital Theatre+
https://www.digitaltheatreplus.com/
Licensing Options: subscription
Content: British theatrical productions, interviews with cast and crew, and documentaries about the production. Includes productions from Shakespeare to contemporary plays, with teaching guides

Provider: Journal of Visualized Experiments (JoVE)
https://www.jove.com/journal
Licensing Options: subscription by subject area
Content: peer-reviewed video research articles and video resources for science education

Provider: Naxos Video Library
https://www.naxosmusiclibrary.com/login
Licensing Options: subscription
Content: filmed performances of classical music, opera, ballet, as well as documentaries and master classes

Provider: OnTheBoards.tv
https://www.ontheboards.tv/
Licensing Options: subscription
Content: contemporary performance films and artistic projects

Provider: Psychotherapy.net
https://www.psychotherapy.net
Licensing Options: subscription
Content: psychotherapy video collections that include demonstrations and counseling sessions. Pricing by the number of videos in a collection

Provider: Sage Knowledge
https://learningresources.sagepub.com/video
Licensing Options: subscription, EBA
Content: discipline-specific video collections teaching and research-oriented video within the social sciences (e.g., tutorials, case studies, expert interviews)

Additional Resources

Books

DeCesare, Julie A. 2014. *Streaming Video Resources for Teaching, Learning, and Research*. Chicago: American Library Association.

Duncan, Cheryl J. and Erika Day Peterson. 2014. *Creating a Streaming Video Collection for Your Library*. Lanham, MD: Roman & Littlefield.

Handman, Gary. 2002. *Video Collection Development in Multi-Type Libraries: A Handbook*. Westport, CT: Greenwood Press.

Hartnett, Eric, ed. 2019. *Guide to Streaming Video Acquisitions*. Chicago: American Library Association.

Higgins, Colin. 2015. *Cataloging and Managing Film and Video Collections: A Guide to Using RDA and MARC21*. Chicago: American Library Association.

Hobbs, Renee, Liz Deslauriers, and Pam Steager. 2019. *The Library Screen Scene: Film and Media Literacy in Schools, Colleges, and Communities*. New York: Oxford University Press.

Labato, Ramon. 2019. *Netflix Nations: The Geography of Digital Distribution*. New York: New York University Press.

Laskowski, Mary S. 2011. *Guide to Video Acquisitions in Libraries: Issues and Best Practices*. Chicago: American Library Association.

Spina, Carli. 2021. *Video Accessibility*. Library Technology Reports 57(3). Chicago: American Library Association.

Articles

Adamich, Tom. 2021. "ON-DEMAND STREAMING MEDIA-Considerations for Public Libraries." *Computers in Libraries* 41 (10): 18–21.

Adams, Tina M. and Claudia C Holland. 2018. "Streaming Media in an Uncertain Legal Environment: A Model Policy and Best Practices for Academic Libraries." *Journal of Copyright in Education & Librarianship* 1 (2). https://doi.org/10.17161/jcel.v1i2.6550.

Alvin, Glenda. 2013. "How to Engage Faculty with Academic Video." *Against the Grain* 25: 56.

Aravopoulou, Eleni, Merlin Stone, and Luisa Weinzierl. 2017. "Modernising the Curriculum and Pedagogy–to Be or Not to Be? Using Film and Online Video to Engage Students and Enhance Learning." *International Journal of Higher Education Management* 4 (1): 1–18. https://research.stmarys.ac.uk/id/eprint/1732/.

Ariew, Susan, LeEtta Schmidt, and Matt Torrence. 2021. "A Guide to Openly Accessible Media and Streaming Video Content." *Choice: Current Reviews for Academic Libraries* 58 (5): 433–40.

Association of College & Research Libraries. 2020. "2020 Top Trends in Academic Libraries: A Review of the Trends and Issues Affecting Academic Libraries in Higher Education." *College & Research Library News* 81 (6): 270–8. https://doi.org/10.5860/crln.81.6.270.

Austin, Cathy and Bonnie Thornton. 2022. "Streaming Video and Academic Libraries: A Survey on Commonalities in Acquisitions." *Journal of Electronic Resources Librarianship* 34 (3): 245–55. https://doi.org/10.1080/1941126X.2022.2099040.

Backowski, Roxanne. 2021. "Collaborations between Libraries and Instructional Designers: A Lifeline for Supporting Streaming Video and COVID-19 Induced Changes on Campus." *Journal of Electronic Resources Librarianship* 33 (2): 116–18. https://doi.org/10.1080/1941126X.2021.1912542.

Beisler, Amalia, Rosalind Bucy, and Ann Medaille. 2019. "Streaming Video Database Features: What Do Faculty and Students Really Want?" *Journal of Electronic Resources Librarianship* 31 (1): 14–30. https://doi.org/10.1080/1941126X.2018.1562602.

Benedetti, Susannah. 2016. "Faculty and Student Engagement with Streaming Video." *Against the Grain* 28: 53.

Bergman, Barbara J. 2010. "Making the Most of Your Video Collection: Trends in Patron Access and Resource Sharing." *Library Trends* 58 (3): 335–48. https://doi.org/10.1353/lib.0.0096.

Blick, William. 2021. "Streaming Films and Online Learning during Covid-19 and Onward." *Journal of Electronic Resources Librarianship* 33 (2): 119–20. https://doi.org/10.1080/1941126X.2021.1912543.

Breeding, Marshall. 2022. "Benefits and Challenges for Streaming Media in Libraries." *Computers in Libraries* 42 (4): 9–11.

Bulock, Chris. 2020. "Temporary Free Access." *Serials Review* 46 (4): 305–7. https://doi.org/10.1080/00987913.2020.1850042.

Cagle, Chris. 2019. "Kanopy: Not Just Like Netflix, and Not Free." *Film Quarterly Quorum*, May. https://filmquarterly.org/2019/05/03/kanopy-not-just-like-netflix-and-not-free/.

Carr, Patrick. 2002. "The Development of Feature Film Collections in Academic Libraries." *Current Studies in Librarianship* 26 (1-2): 39–51.

Clark, Joe C., Jonathan Sauceda, and Sheridan Stormes. 2019. "Faculty Format Preferences in the Performing Arts: A Multi-Institutional Study." *College & Research Libraries* 80 (4): 450–69.

Coleman, Nancy. 2019. "New York City's Public Libraries to End Film Streaming Through Kanopy." *New York Times*, June. https://www.nytimes.com/2019/06/24/movies/new-york-citys-libraries-drop-kanopy-streaming.html.

Cooper, Danielle. 2021. "Making Streaming Media Sustainable for Academic Libraries: Launching a New Cohort-Based Collaborative Project." New York: Ithaka S + R. .

Cooper, Danielle and Dylan Ruediger. 2021. "Working with Libraries to Navigate the Streaming Media Environment." New York: Ithaka S + R. https://sr.ithaka.org/blog/workingwith-libraries-to-navigate-the-streaming-media-environment/.

Cooper, Danielle and Rodger Schonfield. 2021. "A Media Empire for the (Public) Library? OverDrive Acquires Kanopy." *The Scholarly Kitchen*, June 15, 2021.

Cross, Cheryl, Christine Fischer, and Cathy Rothermel. 2014. "Streaming Film: How to Serve Our Users." *Serials Review* 40 (3): 154–7.

Cross, Jeanne G. and Susannah J. Benedetti. 2023. "Narrative Films in an Academic Library: How Faculty Requests Inform Collection Development." *Collection Management* 48 (2): 114–26.

Cross, William. 2016. "More than a House of Cards: Developing a Firm Foundation for Streaming Media and Consumer-Licensed Content in the Library." *Journal of Copyright in Education & Librarianship* 1 (1): 1–24. https://doi.org/10.17161/jcel.v1i1.5919.

DeCesare, Julie A. 2014. "Patron-Driven Access to Streaming Video: Profile of Kanopy Streaming." *Against the Grain* 26 (4). https://doi.org/10.7771/2380-176X.6906.

Derry, Sebastian. 2009. "Biz of Acq: Shooting the Rapids — Navigating Changing Video Formats." *Against the Grain* 21 (1). https://doi.org/10.7771/2380-176X.2520.

DiVittorio, Katy, Philip Gaddis, Sommer Browning, Molly Rainard, and Charissa Brammer. 2019. "SILLVR: Streaming Interlibrary Loan Video Resources." Collaborative Librarianship 11 (4): 295–312. https://doi.org/10.1080/0361526X.2022.2040234.

Dixon, Jennifer A. 2017. "The Academic Mainstream: Streaming Video is Becoming a Mainstay on College Campuses, but Discovery and Walled-off Content Create Their Own Challenges." *Library Journal* 142 (14): 42–5.

Doering, David. 2001. "Recipes for Successful Streaming Video." *EMedia Magazine* 14 (3): 45.

Dotson, Daniel S. 2023. "Streaming Videos: Subject Area Usage Trends." *Collection Management* 48 (2): 97–113. https://doi.org/10.1080/01462679.2022.2081831.

Dunn, Cathy. 2020. "Streaming Video Acquisitions: Factors to Consider." *Journal of Electronic Resources Librarianship* 32 (2): 133–5. https://doi.org/10.1080/1941126X.2020.1739852.

Eng, Sidney and Flor A. Hernandez. 2006. "Managing Streaming Video." *Library Collections, Acquisitions, & Technical Services* 30 (3–4): 214–23. https://doi.org/10.1080/14649055.2006.10766129.

Enis, Matt. 2015. "On Demand: Streaming Technology Is Revitalizing Video as an Educational Tool in Academia, but there are Challenges Ahead for Libraries." *Library Journal* 140 (17): 45–9.

Enis, Matt. 2021. "Full Stream Ahead: The Popularity of Public Library Streaming Media Entertainment Spiked during the Pandemic. What's Next?" *Library Journal* 146 (6): 76.

Falloon, Kerry A. 2020. "Evaluating PDA Using Financial Analysis for Streaming Videos: A Case Study at a US Academic Library." *Collection & Curation* 39 (2): 47–52. https://doi.org/10.1108/CC-03-2019-0008.

farrelly, deg. 2016. "Digital Video: Merrily Merrily, Merrily, Merrily, Wading into the Stream: Freed from the Limitations of a Physical Artifact, Streaming Videos Are Available Anywhere, Anytime. No Need to Reserve in Advance, Arrange for Equipment, or Pick up and Return." *Computers in Libraries* 36 (9): 4–8.

farrelly, deg and Jane Hutchison. 2014. "ATG Special Report: Academic Library Streaming Video: Key Findings from the National Survey." *Against the Grain* 26 (5). https://doi.org/10.7771/2380-176X.6852.

Finlay, S. Craig, Michael Johnson, and Cody Behles. 2014. "Streaming Availability and Library Circulation: An Exploratory Study." *LIBRES: Library & Information Science Research Electronic Journal* 24 (1): 1–10.

Foley, Marianne. 2021. "Trends in Streaming Video Usage at a University during the COVID-19 Pandemic." *Journal of Electronic Resources Librarianship* 33 (4): 252–61. https://doi.org/10.1080/1941126X.2021.1988449.

Foley, Marianne. 2023. "Evaluation of Streaming Video Usage in a University Library before, during, and after the COVID-19 Pandemic Lockdown." *Journal of Academic Librarianship* 49 (3): 102698. https://doi.org/10.1016/j.acalib.2023.102698.

Forstot-Burke, Corinne. 2019. "Turn Down for What: A Study of Physical and Streamed Media Usage at the University of Kansas Libraries." *Music Reference Services Quarterly* 22 (4): 189–208. https://doi.org/10.1080/10588167.2019.1570439.

Foster, Anita K. and Gene R. Springs. 2022. "Running up the Hill – Long-Term Streaming Video Pilots: Process, Analysis and Outcomes." *Collection & Curation* 41 (2): 37–46. https://doi.org/10.1108/CC-12-2020-0046.

Fruin, Christine. 2012. "Struggles and Solutions for Streaming Video in the Online Classroom." *American Journal of Distance Education* 26 (4): 249–59. https://doi.org/10.1080/08923647.2012.728078.

Fullmer, Millicent. 2023. "Future Proofing Streaming Video Acquisitions: A Medium Sized Academic Library Adapts." *Technical Services Quarterly* 40 (2): 59–75. https://doi.org/10.1080/07317131.2023.2187109.

Gilson, Tom and Katina Strauch. 2021. "ATG Interviews David Parker." *Against the Grain* 33 (2): 56–7.

Grombly, Amanda A. 2020. "Streaming Video Acquisition Models for Small to Medium Academic Libraries." *The Serials Librarian* 79 (1–2): 163–76. https://doi.org/10.1080/0361526X.2020.1772169.

Grove, Tara Murray. 2021. "Academic Library Video Services: Charting a Post-COVID Course." *Pennsylvania Libraries: Research & Practice* 9 (2): 101–10. https://doi.org/10.5195/palrap.2021.262.

Handman, Gary. 2010. "License to Look: Evolving Models for Library Video Acquisition and Access." *Library Trends* 58 (3): 324–34. https://doi.org/10.1353/lib.0.0094.

Harrington, Patrick. 2019. "Generating Geographic Terms for Streaming Videos Using Python: A Comparative Analysis." *The Code4Lib Journal*, no. 45 (August). https://journal.code4lib.org/articles/14676.

Healy, Ciara. 2010. "Netflix in an Academic Library: A Personal Case Study." *Library Trends* 58 (3): 402–11.

Hilburn, Jessica. 2021. "OVERDRIVE ACQUIRES KANOPY, WHILE AMAZON DEALS WITH DPLA: The Ethical Impacts of Digital Library Resource Consolidation." *Information Today* 38 (7): 16.

Hill, Kate and Neah Ingram-Monteiro. 2021. "What Patrons Really Want (In Their Streaming Media): Using Focus Groups to Better Understand Emerging Collections Use." *Serials Librarian* 80 (1–4): 30–33.

Horbal, Andy. 2018. "Instructor Use of Educational Streaming Video Resources." *The Journal of Academic Librarianship* 44 (2): 179–89. https://doi.org/10.1016/j.acalib.2018.02.009.

Kang, Jennifer M. and Amanda D. Lotz. 2024. "Relocating Video Cultures." *International Journal of Cultural Studies* 27 (1): 3–8. https://doi.org/10.1177/13678779231202545.

King, Rachel. 2014. "House of Cards: The Academic Library Media Center in the Era of Streaming Video." *The Serials Librarian* 67 (3): 289–306. https://doi.org/10.1080/0361526X.2014.948699.

King, Rachel P. 2016. "Access to Circulating Videos in Academic Libraries: From Policy Review to Action Plan." *Collection Management* 41 (4): 209–20.

Knab, Sheryl, Tom Humphrey, and Caryl Ward. 2016. "Now Streaming: A Consortial PDA Video Pilot Project." *Collaborative Librarianship* 8 (1): 41–54.

Koennecke, Jesse. 2015. "Finding Solutions for Streaming Video at Cornell University Library." *Against the Grain* 27 (3): 58–9.

Kolendo, Joanna, Azungwe Kwembe, and Charlene Snelling. n.d. "Piloting the Surge: Streaming Video and Academic Libraries." http://dx.doi.org/10.5703/1288284317160.

Kreamer, Jean Thibodeaux. 1994. "Video and Libraries: Present Status and Future Trends." *The Acquisitions Librarian* 6 (11): 19–28. https://doi.org/10.1300/J101v06n11_03.

Laskowski, Mary S. 2004. "Stop the Technology, I Want to Get Off." *The Acquisitions Librarian* 16 (31–32): 217–25. https://doi.org/10.1300/J101v16n31_17.

Laskowski, Mary S. and Thomas H. Teper. 2014. "Promoting Use and Preserving Access: Navigating the Evolving Nature of Academic Media Collections." *Library Collections, Acquisitions, & Technical Services* 38 (3–4): 54–62. https://doi.org/10.1080/14649055.2015.1055999.

Lear, Christian. 2022. "Controlled Digital Lending of Video Resources: Ensuring the Provision of Streaming Access to Videos for Pedagogical Purposes in Academic Libraries." *Journal of Copyright in Education and Librarianship* 5 (1). https://doi.org/10.17161/jcel.v5i1.14807.

Leffler, Jennifer, Jessica Hayden, and Todd Enoch. 2017. "Juggling a New Format with Existing Tools: Incorporating Streaming Video into Technical Services Workflows." *The Serials Librarian* 72 (1–4): 102–4. https://doi.org/10.1080/0361526X.2017.1284498.

Levenson, Helen N. and Shawn V. Lombardo. 2023. "The State of Streaming Video Content at Academic Libraries." *Collection Management* 48 (1): 1–12. https://doi.org/10.1080/01462679.2023.2255561.

Lobato, Ramon and Amanda D. Lotz. 2020. "Imagining Global Video: The Challenge of Netflix." *JCMS: Journal of Cinema and Media Studies* 59 (3): 132–6. https://doi.org/10.1353/cj.2020.0034.

Lohmann, Sam and Linda Frederiksen. 2018. "Faculty Awareness and Perception of Streaming Video for Teaching." *Collection Management* 43 (2): 101–19. https://doi.org/10.1080/01462679.2017.1382411.

Lowe, Randall A., Rick Davis, Leigh Ann DePope, Deborah K. Li, Robert K. Reeves, and Cindy Steinhoff. 2020. "Managing Streaming Video: The Experiences of Six Academic Libraries." *Journal of Electronic Resources Librarianship* 32 (2): 119–26. https://doi.org/10.1080/1941126X.2020.1739831.

Lupton, Aaron, Ian Gordon, and Helen Salmon. 2018. "Catching the Next Wave: Streaming Video in Academic Libraries." *Open Shelf*, December 3. https://open-shelf.ca/181201ocula-streaming-video/.

Mason, Katherine (Kathie) F. 2022. "Managing Streaming Video License Renewals." *Serials Review* 48 (3–4): 206–7. https://doi.org/10.1080/00987913.2022.2097849.

McAdam, Susan. 1999. "Selection/Acquisition of Non-Book Materials." *The Acquisitions Librarian* 11 (21): 125–45. https://doi.org/10.1300/J101v11n21_11.

McGeary, Bryan James. 2015. "Accessibility, Collaboration, and Staffing: Revamping the Model for Academic Library Video Collections." *Public Services Quarterly* 11 (4): 307–18. https://doi.org/10.1080/15228959.2015.1095672.

McKenzie, Rue and LeEtta M. Schmidt. 2012. "Isn't Everything Online Yet? Streaming Media and Electronic Reserves." *Journal of Interlibrary Loan, Document Delivery & Electronic Reserve* 22 (3–4): 175–80. https://doi.org/10.1080/1072303X.2012.732923.

Miller, Sandra L. 2013. "Innovating to Meet the Demand for Streaming Video." *Journal of Library Innovation* 4 (1): 29–43.

Montavon-Green, Jen. "'What Do I Do With This Video File?': Considerations for Choosing a Video Hosting Platform." *Internet Reference Services Quarterly* (2024): 1–3. 10.1080/10875301.2024.2339878.

Morris, Sara E. and Lea H. Currie. 2016. "To Stream or Not to Stream?" *New Library World* 117 (7/8): 485–98.

Murray Grove, Tara. 2021. "Academic Library Video Services: Charting a Post-Covid Course." *Pennsylvania Libraries* 9 (2): 101–10.

Norton, Brynne. 2022. "Streaming Media Reserves: A Survival Guide." *Journal of Access Services* 19 (4): 111–26. https://doi.org/10.1080/15367967.2022.2131435.

Otto, Jane Johnson. 2014. "University Faculty Describe Their Use of Moving Images in Teaching and Learning and Their Perceptions of the Library's Role in That Use." *College & Research Libraries* 75 (2): 115–44. https://doi.org/10.5860/crl12-399.

Prelitz, Keri. 2022. "Revisiting the Streaming Video Patron-Driven Acquisition Program." *Journal of Electronic Resources Librarianship* 34 (4): 323–31. https://doi.org/10.1080/1941126X.2022.2133919.

Prelitz, Keri and Ann Roll. 2021. "Ten Years of Demand-Driven Acquisition." *Against the Grain* 33 (4): 12–14.

Proctor, Julia. 2018. "Testing Assumptions: Is Streaming Video Really Preferable to DVDs?" *Journal of Electronic Resources Librarianship* 30 (2): 84–90. https://doi.org/10.1080/1941126X.2018.1465516.

Prosser, Helen Gail. 2006. "Video Streaming in the Wild West." *Partnership: The Canadian Journal of Library and Information Practice and Research* 1 (1): 1–9. https://doi.org/10.21083/partnership.v1i1.105.

Rodgers, Andrea. 2019. "Once upon a Time in Streaming Video: A Community College's Adventure with Kanopy's PDA Model." *College & Research Libraries News* 80 (9): 498. https://doi:10.5860/crln.80.8.498.

Rodgers, Wendy. 2018. "Buy, Borrow, or Steal? Film Access for Film Studies Students." *College & Research Libraries* 79 (4): 568–91. https://doi.org/10.5860/crl.79.4.568.

Rodriguez, Michael. 2021. "Flipping Subscription and Access Fees to Perpetuity." *Serials Review* 47 (3/4): 187–90. https://doi.org/10.1080/00987913.2021.2017826.

Roll, Ann. 2019. "Report of the Session 'The State of Streaming Video in Academic Libraries,' Co-Sponsored by the ALCTS Collection Management Section (CMS) Collection Development Librarians of Academic Libraries Interest Group and the Film and Media Round Table. American Library Association Midwinter Meeting, Seattle, January 2019." *Technical Services Quarterly* 36 (4): 408–10. https://doi.org/10.1080/07317131.2019.1664109.

Russell, Carrie. 2010. "The Best of Copyright and VideoLib." *Library Trends* 58 (3): 349–57. https://doi.org/10.1353/lib.0.0095.

Schroeder, Rebecca and Julie Williamsen. 2011. "Streaming Video: The Collaborative Convergence of Technical Services, Collection Development, and Information Technology in the Academic Library." *Collection Management* 36 (2): 89–106.

Serrano, Amauri and Michael Fernandez. 2023. "Impacts of COVID-19: Toward a Streaming-Preferred Video Collection Policy." *Collection Management* 48 (1): 5–21. https://doi.org/10.1080/01462679.2022.2124392.

Singh, Kelly. 2020. "The Value of Video: Accessibility, Streaming, and the 21st Century Library." *Against the Grain* 32 (3): 59.

Spicer, Scott and Andrew Horbal. 2017. "The Future of Video Playback Capability in College and University Classrooms." *College & Research Libraries* 78 (5): 706–22. https://doi.org/10.5860/crl.78.5.706.

Streeter, Abigail, Kimberly Kennedy, and Jamie Saragossi. 2023. "Streamlining Streaming Videos: Moving towards an Efficient Streaming Video Workflow in Alma/Primo." *Journal of Electronic Resources* Librarianship 35 (3): 171–81. https://doi.org/10.1080/1941126X.2023.2224663.

Thibeault, Jason. 2020. "Streaming Video Fundamentals." *SMPTE Motion Imaging Journal* 129 (3): 10–15.

Towery, Stephanie S., Amanda N. Price, and Karen E. Cowen. 2019. "Video Streaming Licenses: Using a Decision Tree and Workflow Chart." *Journal of Copyright in Education & Librarianship* 3 (1). https://doi.org/10.17161/jcel.v3i1.7483.

Trujillo, Nicole, Erik Radio, and Melanie Walker. 2020. "What Metadata Matters?: Correlation of Metadata Elements with Click-Through Rates for E-Books and Streaming Video in the Academic Library Catalog." *Journal of Web Librarianship* 14 (3–4): 86–99. https://doi.org/10.1080/19322909.2020.1850390.

Urban, Sandra Gall. 2022. "Evaluating Kanopy Access Models in Academic Libraries: Balancing Demand and Budget Constraints." *Collection Management* 47 (1): 37–48. https://doi.org/10.1080/01462679.2021.1940410.

Vallier, John. 2010. "Twenty-First Century Academic Media Center: Killer App or Chindogu?" *Library Trends* 58 (3): 378–90.

VanUllen, Mary K., Emily Mock, and Emmalyn Rogers. 2018. "Streaming Video at the University at Albany Libraries." *Collection & Curation* 37 (1): 26–9. https://doi.org/10.1108/NLW-03-2016-0021.

Wahl, Mary. 2017. "Full Stream Ahead: Designing a Collection Development Workflow for Streaming Video Content." *Library Resources & Technical Services* 61 (4): 226–36.

Walters, William H. 2003. "Video Media Acquisitions in a College Library." *Library Resources & Technical Services* 47 (4): 160–70. https://doi.org/10.5860/lrts.47n4.160.

Wang, Jian and Elsa Loftis. 2020. "The Library Has Infinite Streaming Content, but Are Users Infinitely Content? The Library Catalog vs. Vendor Platform Discovery." *Journal of Electronic Resources Librarianship* 32 (2): 71–86. https://doi.org/10.1080/1941126X.2020.1739818.

Widzinski, Lori. 2010. "'Step Away from the Machine': A Look at Our Collective Past." *Current Trends in Academic Media Collections and Services* 58 (3): 358–77.

Willson-Metzger, Alicia. 2018. "Don't Let's Ask for the Moon, We Have the Stars: Starting a Streaming Video Program on a Limited Budget." *Against the Grain* 30 (4): 67.

Yi, Hua. 2019. "Developing a Media Collection That Addresses Trends While Targeting the Needs of Users: An Analysis of Online Streaming Videos (OSV) at CSU San Marcos." *International Journal of Librarianship* 4 (1): 59–71.

Book Chapters

farrelly, deg. 2014. "Streaming Video." In Becky Albitz, Diane Zabel, and Christine Avery (eds.), *Rethinking Collection Development and Management*, 215–29. Englewood: ABC-CLIO, LLC.

Paulson, Kari. 2011. "The Story of Patron-Driven Acquisition." In David A. Swords (ed.), *Patron-Driven Acquisitions*, 63–78. Berlin, Boston: De Gruyter Saur. https://doi.org/10.1515/9783110253030.63.

Shelton, Trey. 2016. "Streaming Video." In *Implementing and Accessing Use-Driven Acquisitions: A Practical Guide for Librarians*, 108–21. Practical Guides for Librarians 23. Lanham: Rowan & Littlefield Publishers.

Spicer, Scott. 2018. "Perspectives on the role of instructional video in higher education." In *The Routledge Companion to Media Education, Copyright, and Fair Use*, 236–57. New York: Routledge.

White Papers and Reports

Bossenga, Susie, Chris Bulock, Luann DeGreve, Carol Doyle, Tom Goetz, Ruth Lindemann, Stephen McMinn and Charles Uth. 2013. "Streaming Video in Academic Libraries." *Commercial Products Committee White Paper*. Consortia for Academic and Research Libraries (CARLI). https://www.carli.illinois.edu/sites/files/files/2014CommercialProductsCommStreamingVideoinAcademicLibraries.pdf.

Carmichael, Michael, Abigail-Kate Reid, and Jeffrey D Karpicke. 2018. "Assessing the Impact of Educational Video on Student Engagement, Critical Thinking and Learning: The Current State of Play." White Paper. Sage. https://group.sagepub.com/white-paper-archive/assessing-the-impact-of-educational-video-on-student-engagement-critical-thinking-and-learning-the-current-state-of-play.

Cooper, Danielle M., Dylan Ruediger, and Makala Skinner. 2022. *Streaming Media Licensing and Purchasing Practices at Academic Libraries: Survey Results*. Ithaka S+R, June 9. https://doi.org/10.18665/sr.316793.

Kanopy. 2021. "Streaming Video Trends in Academic Libraries." White Paper. Kanopy. https://storage.pardot.com/54092/1615316161w8o0lP4M/Academic_Library_Survey_Whitepaper_3_9_21.pdf.

Kaufman, Peter B. and Jen Mohan. 2009. "Video Use and Higher Education: Options for the Future." Intelligent Television. https://intelligenttelevision.com/files/42-intcccnyuvideo_and_higher_edjune_2009_2.pdf.

Leonard, Elizabeth. 2015. "Great Expectations: Students and Video in Higher Education." A SAGE White Paper. https://us.sagepub.com/sites/default/files/studentsandvideo.pdf.

MacDougall, Ruby and Dylan Ruediger. 2023. *Teaching with Streaming Video: Understanding Instructional Practices, Challenges, and Support Needs. Ithaka S+R*. January 26. https://doi.org/10.18665/sr.318216.

Tanasse, Gisèle. 2021. "Implementing and Managing Streaming Media Services in Academic Libraries." *Choice 360 White Paper*. Association of College and Research Libraries. https://www.choice360.org/research/implementing-and-managing-streaming-mediaservices-in-academic-libraries/.

Presentations and Conference Proceedings

Barker, Cara, Whitney Jordan, and Jessica Zellers. 2018. "Starting a Streaming Video Program on a Limited Budget." *Proceedings of the Charleston Library Conference.* https://doi.org/10.5703/1288284316675.

Calabrese, Cara Mia, and Michael A Arthur. 2021. "Expectations and Realities of Streaming Video." Presentation at the Association of College and Research Libraries Conference, Seattle, WA. http://ir.ua.edu/handle/123456789/7495.

Deck, Trevor. 2020. "Streaming Media and Unsustainable Cost Models: Impacts on Libraries." Presentation at the American Library Association Midwinter Meeting, Philadelphia, PA.

Denzer, Kelly. 2023. "A Furry Brown Bear Sits in the Snow, its Fur Speckled with White: Audio Description in the Wild." Poster Presentation at the Electronic Resources and Libraries Conference, Austin, TX.

Dewitt-Miller, Erin, Susan Smith, Xin Wang, Allyson Rodrigues, and Emily Billings. 2017. "How do I Know if it's useful if I can't even get it to Open? Assessing Information Interaction at the University of North Texas to Improve Library Collections and Services." Presentation at the Electronic Resources and Libraries Conference, Austin, TX. https://digital.library.unt.edu/ark:/67531/metadc977349/.

Duncan, Cheri and Erika Peterson. 2012. "You Ought to Be in Pictures: Bringing Streaming Video to Your Library." *Proceedings of the Charleston Library Conference*. https://doi.org/10.5703/1288284314946.

Falloon, Kerry. 2017. "Rolling with PDA and DDA: How Academic Libraries Can Use Patron-Driven and Demand-Driven Acquisition Techniques to Build Library Collections With Minimal Management and Budget." *Proceedings of the Charleston Library Conference*. https://doi.org/10.5703/1288284316455.

farrelly, deg. 2013. "Freely Flowing: Openly Accessible Sources for Streaming Video." *Proceedings of the Charleston Library Conference*. https://doi.org/10.5703/1288284315140.

Foley, Marianne. 2019. "Streaming Video PDA: Brace Yourself, Usage is Coming." *Proceedings of the Charleston Library Conference*. https://doi.org/10.5703/1288284317017.

Forstot-Burke, Corinne and Lea Currie. 2019. "What Makes Us Do It? The Legalities and Demand That Necessitate a Library Video Streaming Service." *Proceedings of the Charleston Library Conference*. https://doi.org/10.5703/1288284317040.

Fountain, Kathleen. 2012. "Managing Expectations and Obligations: The Librarian's Role in Streaming Media for Online Education." *Proceedings of the Charleston Library Conference*. https://doi.org/10.5703/1288284314954.

Fruin, Christine, Trey Shelton, Aimee Barrett, and Allison Jai O'Dell. 2017. "'We'll Do It Live': Building Access to Video Content Based on Freedoms of Use." *Proceedings of the Charleston Library Conference*. https://doi.org/10.5703/1288284316453.

Garewal, Kevin and Stephanie Church. 2020. "Navigating Collection Development Issues with Leased Streaming Media." Presentation at the American Library Association Midwinter Meeting, Philadelphia, PA.

Keyes, Joshua and Elsa Loftis. 2019. "Navigating the Sustainable Stream: Academic Libraries Keeping Pace with Streaming Content Demand." Presentation at the

Acquisitions Institute, Timberline Oregon, May 19. https://archives.pdx.edu/ds/psu/28850.
Koennecke, Jesse, Susan Marcin, and Matthew Pavlick. 2015. "The Devil Is in the Details: Managing the Growth of Streaming Media in Library Collections." *Proceedings of the Charleston Library Conference*. https://doi.org/10.5703/1288284315646.
Kolendo, Joanna, Azungwe Kwembe, and Charlene Snelling. 2020. "Piloting the Surge: Streaming Video and Academic Libraries." *Proceedings of the Charleston Library Conference*. https://doi.org/10.5703/1288284317160.
McCleskey, Sarah, Christine Fischer, Steven Milewski, and Jim Davis. 2015. "You've Licensed It. Now What?" Proceedings of the Charleston Library Conference. *Proceedings of the Charleston Library Conference*. https://doi.org/10.5703/1288284315643.
Milewski, Steven and Monique Threatt. 2016. "Apples to Oranges: Comparing Streaming Video Platforms." *Proceedings of the Charleston Library Conference*. https://doi.org/10.5703/1288284316434.
Mower, Allyson, Mary Ann James, Catherine Soehner, Maria Hunt, Dave Heyborne, and Joni Clayton. 2016. "One Library's Successful Venture in Providing Comprehensive Streaming Media Services." *Proceedings of the Charleston Library Conference.* https://doi.org/10.5703/1288284316240.
Serrano, Amauri. 2020. "I Saw It and I Want It: Filling Course-Related Streaming Video Requests." Presentation at the American Library Association Midwinter Meeting, Philadelphia, PA.
Tennant, Shannon. 2020. "Navigating the Transition to Mediated Streaming Video." Presentation at the American Library Association Midwinter Meeting, Philadelphia, PA.
Van Gaasbeck, Kalvin. 2021. "Streaming Video Licensing Workflow Using Google Sheets." Presented at the NASIG Annual Conference, Virtual.
Voelker, Greg, Richard Zwiercan, and Michael Frazier. 2014. "Navigating the Flow of Value Streams to the Seas of Collection Management, Acquisitions, and Preservation." *Proceedings of the Charleston Library Conference*. https://doi.org/10.5703/1288284315266.

Webinars

MacDougall, Ruby. 2023. "Teaching with Streaming Video: Understanding Instructional Practices, Challenges, and Support Needs." Webinar, Association of College and Research Libraries, May 2. https://www.choice360.org/webinars/teaching-with-streaming-video-understanding-instructional-practices-challenges-and-support-needs/.
Machado-Dillon, Christina, Becky Imamoto, Jeff Cola, and Ariel Marbury. 2021. "How Academic Libraries Are Bringing Diversity, Equity and Inclusion to their Campuses." Webinar, Association of College and Research Libraries, October 7. https://www.choice360.org/webinars/how-academic-libraries-are-bringing-diversity-equity-and-inclusion-to-their-campuses/.
Video Trust Coffee Chats - https://www.videotrust.org/coffeechats archives are accessible with a Video Trust membership.

Guidelines

Association of College and Research Libraries. 2018. "Guidelines for Media Resources in Academic Libraries." Chicago: American Library Association. www.ala.org/acrl/standards/mediaresources.

OLAC Cataloging Policy Committee Streaming Media RDA Guide Task Force. 2022. "Best Practices for Cataloging Streaming Media." Online Audiovisual Catalogers Inc. https://cornerstone.lib.mnsu.edu/olac-publications/25/.

Professional Organizations

Film and Media Round Table (FMRT) – American Library Association https://www.ala.org/rt/fmrt

Provides leadership within the American Library Association (ALA) on all issues related to video collections, programs, and services in libraries.

FMRT Professional Resources for Librarians https://www.ala.org/rt/fmrt/professionalresources/vrtresources/resources

Video Trust

Supports and educates library professionals on issues related to visual media acquisition, usage, preservation, and access. Offers individual and institutional membership.

Listservs

Film and Media Round Table Discussion Lists - Available to members of Film and Media Round Table (FMRT) - American Library Association

VideoLib - videolib@lists.berkeley.edu

"List to discuss issues related to the selection, evaluation, acquisition, bibliographic control, preservation, and use of current and evolving video formats in libraries and related institutions."

VideoNews - videonews@lists.berkeley.edu

"Electronic clearinghouse for information about new services, products, resources, and programs of interest to video librarians and archivists, educators, and others involved in the selection, acquisition, programming, and preservation of video materials in non-profit settings."

Catalogs and Discovery Tools (freely available)

AFI (American Film Institute) Catalog

https://aficatalog.afi.com/

Detailed descriptions of American films from 1893-1975, with full or short records for films from 1976-present.

International Federation of Film Archives – Film/AV Collections of FIAF Affiliates Online

https://www.fiafnet.org/pages/E-Resources/Film-AV-Collections-Online.html

List of links to free online streaming services used by FIAF affiliates to provide access to their audiovisual collections.

IMDB (Internet Movie Database)

https://www.imdb.com/

Database of film and television titles with detailed cast and production crew information.

JustWatch

https://www.justwatch.com/

Use to source available streaming versions of films and television series across free and subscription platforms.

National Archives Catalog

https://www.archives.gov/research/motion-pictures

Use to find digitized versions of films held at the National Archives.

Glossary

ADA The Americans with Disabilities Act is a civil rights law in the United States prohibiting discrimination toward individuals with disabilities. Any entity receiving public funding must adhere to the ADA, which includes most libraries.

Audio Descriptions Voiceover describing the visual information onscreen, accessibility aid for users with visual disabilities.

Authorized Users Set of individuals affiliated with an institution licensing content who are granted access and rights of use.

Captions Text of dialogue, sound effects, musical cues, and any other audio information that appears on screen to assist users with auditory disabilities. Captions can be toggled on/off (closed captions) or embedded with the visual (open captions). Distinct from subtitles as they include all audio information, not just translated dialogue.

CDL Controlled digital lending (CDL), a method for the scholarly sharing of digital assets whereby content such as whole e-books or streaming videos are lent from one library to another via a mechanism that is regulated in a manner similar to physical material lending.

Click-through Agreement License model for digital content whereby an end user is prompted to accept all of the licensor's terms and conditions before being granted access. Typically, the user affirms their acceptance by clicking an agree to button in a pop-up window or dialog box.

CMS Content management systems (CMS) are locally hosted or cloud-based systems for centralizing, managing, and delivering digital content to users. Common examples utilized by libraries for managing streaming video content include Kaltura and Panopto.

Consortium A partnership of libraries working toward common goals in areas of collective interest such as licensing, acquisitions, systems, or other shared infrastructure. Typically geographically based, a consortium may operate on a statewide, regional, or national level.

Content Provider General term for any entity or body that makes streaming video content available to libraries.

COUNTER Stands for Counting Online Usage of Networked Electronic Resources. COUNTER is a standardized code of practice for publishers of electronic resource content to compile and package reporting of usage.

CPU Cost-per-use, an assessment measure for determining the value of paid content whereby the amount paid is divided by the total usage over a period of time, arriving at a calculated value for a single use.

DDA/PDA Demand-driven acquisitions, also referred to as patron-driven acquisitions, is a method of acquisition in which content is not purchased unless it is used.

DEI Diversity, equity, and inclusion; policies and measures for addressing social justice and ensuring that previously marginalized voices and perspectives are represented. Within streaming video collections can apply to building collections that highlight a diverse array of subjects and creators, as well as making collections accessible to all users.

DFXP Distribution format exchange profile, a caption file format supported by many streaming platforms, including YouTube. Files are encoded in XML and allow for formatting features of captions such as font and location specifications.

Discovery Layer A user-facing tool that allows for searching of library content, typically through a single search box interface similar to internet search engines like Google.

Distributor An entity with the right to supply or make available streaming video content; not necessarily the creator or original rights holder.

DSL A digital site license allows an institution to host a digital file on an internal server or third party for secured access to authenticated users.

EBA Evidence-based acquisitions is a method of acquisition that allows library access to online content for a pre-defined period of time. Once the time period is up, the library makes purchasing decisions based on usage or other selection criteria.

ERMS Electronic resources management systems are internal or third-party databases that allow for triaging information relevant to a library's electronic resource holdings. They can include information on license terms, acquisitions data, usage statistics, or vendor contact information. Will frequently integrate with e-resource knowledge base holdings.

Fair Use Section of US copyright law that permits certain rights for the use of copyrighted intellectual property. For streaming video content licensed by a library, not-for-profit classroom viewing of the licensed content is considered an educational purpose covered under fair use.

First Sale Doctrine Aspect of US copyright law allowing libraries to loan physical material that is purchased and owned, such as books or DVDs. The application of the doctrine to digital content has not been demonstrated conclusively, preventing libraries in most cases from lending electronic materials such as e-books.

ILL Interlibrary loan, a cooperative process for scholarly sharing whereby libraries lend out materials from their held collections to partner libraries.

ILS Integrated library systems are back end management systems that incorporate major aspects of library functionality including cataloging of materials with a public interface for users, acquisitions and budget maintenance, circulation of physical materials, and authorized user identity management.

IP Range Internet protocol range; a common method of mediating access to licensed electronic resource content, whereby only specific addresses or ranges of networked user connections are authorized.

IR Institutional repository, a digital archive of locally held content, frequently highlighting the scholarly output of a college or university. IRs are also used as hosting and publishing platforms.

Just-in-Case Acquisition Proactive model of collection development, where content is acquired in anticipation of future user needs.

Just-in-Time Acquisition Reactive model of collection development, where content is acquired in response to a present user need. Forms of this model include demand-driven acquisition or patron-driven acquisition.

JustWatch A freely available online database that indexes the availability of individual film titles on streaming platforms. Frequently used in streaming video collection development to source title availability.

KB Knowledge bases are databases of available electronic resource titles and packages that can be used to track the active holdings of a library's online content. Frequently made available through third-party service providers and integrated with access management tools such as discovery layers and link resolvers.

License Contract between a library and a content provider that outlines terms of use, user rights, and length of entitlement for acquired content.

Licensee Party in a license agreement who is being granted rights to content.

Licensor Party in a license agreement who holds rights to the content being licensed.

Life of File A common term for perpetually licensed streaming content whereby the library's entitlement persists so long as the digital media file is viable.

Link Resolver Access tool that provides direct linking to full-text e-resource content held by a library and activated through a knowledge base. Content is surfaced by users in the library's discovery layer and content-specific metadata is embedded into links using the OpenURL standard, redirecting users to the article or chapter on the provider platform.

LMS A learning management system is a software, frequently cloud-based, that supports learning and instruction by allowing users to access pedagogical content such as syllabi, course reserves, recorded lecturers, and other instructor-provided materials. Examples of enterprise solutions utilized by educational institutions include Blackboard and Canvas.

Local Hosting/Management Process whereby a library receives a source file from a vendor or provider that in turn is uploaded to a content management system whereby authorized users can gain access.

LSP Library services platforms are back end management systems specifically designed to accommodate electronic and digital formats that increasingly comprise library collections. They differ from predecessor integrated library systems (ILS) that were focused on supporting physical material workflows. Sometimes referred to as next-generation or unified resource management systems, LSPs are cloud-based and do not require the client software of the ILS.

MARC Machine-readable cataloging, a standard for metadata records that are loaded into library online catalogs and other discovery tools.

Mediated Purchase Content is purchased through manual librarian intervention.

MOV A common video container file format for streaming. Generally, MOV files are less compressed and have larger file sizes than MP4 files. They are better suited as a preservation format than MP4 and are designated as a preferred format for video by the Library of Congress.

MP4 MPEG-4, a common video container file format for streaming. Generally, MP4 files are highly compressed and have smaller file sizes than formats with lossless compression. Commonly used for local hosting by libraries because the smaller file sizes can be more quickly downloaded and uploaded to library-managed streaming platforms.

OA Open access is the principle that digital content be made freely available to all users who have an internet connection.

OCLC Library services organization, notable for maintaining the global catalog of library holdings, WorldCat. Additional services provided include ILL, delivery of MARC records, and SaaS products such as the WorldShare Management Services platform and the WorldCat Discovery search layer.

OLAC Online Audiovisual Catalogers, a membership organization that provides networking opportunities and guidelines for library workers engaged in the creation, maintenance, and discovery of metadata for non-print materials, including video and sound recordings, websites, multimedia, and video games.

OPAC Online public access catalog, a user-facing tool for discovering content held by the library.

PAL A perpetual access license provides perpetual access to resources hosted on a third-party platform.

Pareto Principle (80/20 Rule) A statistical concept that theorizes a disproportionate minority (20 percent) of actors is responsible for the large majority (80 percent) of results. An applied example for library materials would be that 80 percent of a library collection's usage comes from 20 percent of titles within that collection.

PO A purchase order is a uniquely identifiable tracking entry for a library acquisition. Includes title information for the acquired content, funding and allocation information for processing payment, and a reference number to share with content providers.

PPR Public performance rights are the permissions granted by the rights holder for a film screening that is made open to the public. The specific terms of the PPR can vary widely, for example, a one-time-only screening, and caps on audience size.

Proxy Server Access method that allows for remote use of library-licensed content by having authorized users authenticate via a whitelisted proxy IP address. OCLC's EZproxy is the most commonly used proxy tool for libraries.

Rights Holder Owner of the copyright for a work of intellectual property.

ROI Return on investment is a ratio for assessment that measures value (the return) in comparison to the amount expended (the investment) over time.

Section 508 Amendment to the US Rehabilitation Act specifying that all electronic and information technology procured with federal funding be made accessible to individuals with disabilities. Commonly invoked for libraries in the United States to make electronic resources accessible.

SERU Shared e-resource understanding, a NISO-developed licensing model whereby libraries and content providers enter into a "handshake" agreement to adhere to a generic set of terms, thereby obviating the need for negotiating and signing a contract. Libraries and providers affirm their participation by joining the SERU Registry.

Sidecar File A closed caption file that is loaded alongside a streaming video and can be toggled on and off-screen.

SRT Subrip subtitle, a common file format for streaming video captions. A simplified format that is timestamped to display onscreen captions in sync with streaming video.

SSO Single sign-on is an authentication method for online access that allows users to authenticate across multiple platforms with a single set of credentials.

Streaming Video Platform A digital repository for the storage and delivery of streaming video content via an online user interface. Can include locally managed platforms, education-focused platforms such as Kanopy, and commercial platforms such as Netflix.

Subtitles Text of dialogue that is typically translated from another language. Subtitles for streaming video content can aid accessibility, though captions are preferred, as subtitles lack cues for sound effects or other auditory information.

Term License A short-term, non-perpetual agreement granting rights of use to content for a specified period of time. For streaming video, this is typically 1- or 3-year increments.

Transcripts Textual documentation of the spoken dialogue for a video. Generally provided as a separate text file from video, though some streaming platforms allow users to toggle a timestamped transcript alongside a playing video. Can aid accessibility, though captions that contain a full account of all auditory cues are preferred.

Unmediated Purchase Content is purchased without librarian intervention.

Vendor In library acquisitions, a general term for any entity that sells physical and/or digital materials to libraries.

Video Trust Membership organization that serves to educate on best practices for the collection and preservation of video content in libraries, as well as to advocate on behalf of the library community and its users.

VIDEOLIB Listserv for communication and discussion on topics relating to the selection, licensing, acquisition, cataloging, access, preservation, and use of video content within libraries and related institutions.

VPAT Voluntary product accessibility template, a document that indicates how content providers comply with accessibility benchmarks such as Section 508 or WCAG.

WCAG The Web Content Accessibility Guidelines are a set of recommendations for making online content more accessible, developed by the World Wide Web Consortium (W3C), an international standards organization. The current recommendation, commonly cited in library licenses as a benchmark for content providers, is WCAG 2.1 Level AA.

WebVTT/VTT Web video text tracks, a common file format for streaming video captions. More complex and robust than SRT, allows for modifications to text formatting and onscreen positioning.

Index